The Effects on the Atmosphere of a Major Nuclear Exchange

Committee on the Atmospheric Effects of Nuclear Explosions

Commission on Physical Sciences, Mathematics, and Resources

National Research Council

NATIONAL ACADEMY PRESS
Washington, D.C. 1985

NATIONAL ACADEMY PRESS 2101 CONSTITUTION AVE., NW WASHINGTON, DC 20418

NOTICE: The project that is the subject of this report was approved by the Governing Board of the National Research Council, whose members are drawn from the councils of the National Academy of Sciences, the National Academy of Engineering, and the Institute of Medicine. The members of the committee responsible for the report were chosen for their special competences and with regard for appropriate balance.

This report has been reviewed by a group other than the authors according to procedures approved by a Report Review Committee consisting of members of the National Academy of Sciences, the National Academy of Engineering, and the Institute of Medicine.

The National Research Council was established by the National Academy of Sciences in 1916 to associate the broad community of science and technology with the Academy's purposes of furthering knowledge and of advising the federal government. The Council operates in accordance with general policies determined by the Academy under the authority of its congressional charter of 1863, which establishes the Academy as a private, nonprofit, self-governing membership corporation. The Council has become the principal operating agency of both the National Academy of Sciences and the National Academy of Engineering in the conduct of their services to the government, the public, and the scientific and engineering communities. It is administered jointly by both Academies and the Institute of Medicine. The National Academy of Engineering and the Institute of Medicine were established in 1964 and 1970, respectively, under the charter of the National Academy of Sciences.

This project was supported by contract DNA001-83-C-0137 between the National Academy of Sciences and the Defense Nuclear Agency, Department of Defense. The Defense Nuclear Agency neither confirms nor denies data cited in this report.

Library of Congress Catalog Number 84-62739
International Standard Book Number 0-309-03528-7

Printed in the United States of America

NATIONAL RESEARCH COUNCIL

2101 CONSTITUTION AVENUE WASHINGTON, D. C. 20418

OFFICE OF THE CHAIRMAN

In early 1983, the Department of Defense asked us to assess information on the possible atmospheric effects of nuclear war. We formed a committee of specialists from relevant fields to conduct the assessment. This is the committee's final report.

Nuclear war would have catastrophic effects beyond those that might degrade the earth's atmosphere; thus our committee examined only one part of a large and complex issue. And even within this part the committee was asked to focus only on effects on the atmosphere and not to carry the analysis to the next logical step: the consequences of changes in the atmosphere for life on earth. This is an issue that should and will be addressed.

The committee has admirably performed a task that proved even more difficult than we had anticipated. We had appreciated the difficulty of examining the scientific aspects of a subject that--for understandable reasons--provokes strong emotional reactions. An equally formidable task, however, was that of coping with profound gaps in existing knowledge. The unfortunate but unavoidable fact is that, even though we are 40 years into the nuclear age, much of the basic information needed to assess the likelihood and extent of global atmospheric consequences of a nuclear exchange simply does not exist. As a result, the committee has been unable to provide the simple, unqualified finding that we all might wish to have in order to assure that any nation's decisions about nuclear forces are not made in ignorance of their true consequences.

Under these conditions the committee determined it could best serve by summarizing existing knowledge, by drawing the partial conclusions (with necessary qualifications) that are supported by data, by clearly describing the nature and extent of uncertainties, and by indicating where those uncertainties might be reduced through further research.

Because additional knowledge might well alter our current understanding, the report can only be viewed as an interim statement. Nevertheless, we believe it can help the scientific community and the world's governments advance the time when we can adequately answer a question of surpassing international importance.

Frank Press
Chairman

Committee on the Atmospheric Effects of Nuclear Explosions

GEORGE F. CARRIER, Division of Applied Sciences, Harvard University, Cambridge, Massachusetts; Chairman
WILLIAM J. MORAN, Vice Admiral USN (Ret.), Los Altos, California; Vice Chairman
JOHN W. BIRKS, Department of Chemistry, University of Colorado, Boulder, Colorado
ROBERT W. DECKER, U.S. Geological Survey, Menlo Park, California
DOUGLAS M. EARDLEY, Institute for Theoretical Physics, University of California, Santa Barbara, California
JAMES P. FRIEND, Department of Chemistry, Drexel University, Philadelphia, Pennsylvania
ERIC M. JONES, Los Alamos National Laboratory, Los Alamos, New Mexico
JONATHAN I. KATZ, Department of Physics, Washington University, St. Louis, Missouri
SPURGEON M. KEENY, JR., National Academy of Sciences, Washington, D.C.
CONWAY B. LEOVY, Department of Atmospheric Sciences and Graduate Program in Geophysics, University of Washington, Seattle, Washington
CONRAD L. LONGMIRE, Mission Research Corporation, Santa Barbara, California
MICHAEL B. McELROY, Harvard Center for Earth and Planetary Physics, Harvard University, Cambridge, Massachusetts
WILLIAM PRESS, Department of Astronomy, Harvard College Observatory, Cambridge, Massachusetts
JACK P. RUINA, Department of Electrical Engineering, Massachusetts Institute of Technology, Cambridge, Massachusetts
EUGENE M. SHOEMAKER, U.S. Geological Survey, Flagstaff, Arizona
LEVERING SMITH, Vice Admiral USN (Ret.), San Diego, California
O. BRIAN TOON, Ames Research Center, NASA, Moffett Field, California
RICHARD P. TURCO, R&D Associates, Marina del Rey, California

Staff

LAWRENCE E. McCRAY
PEGGY POWERS

Commission on Physical Sciences, Mathematics, and Resources

Acknowledgments

The committee expresses its appreciation for the contributions of the following individuals:

Thomas Ackerman, National Aeronautics and Space Administration
Marcia Baker, University of Washington
Robert Cess, State University of New York, Stony Brook
Robert Charlson, University of Washington
Anthony Clarke, University of Washington
Peter Connell, Lawrence Livermore National Laboratory
Curt Covey, National Center for Atmospheric Research
John De Ris, Factory Mutual Corporation
Robert Dickinson, National Center for Atmospheric Research
Frank Fendell, TRW, Inc.
Paul Guthals, Los Alamos National Laboratory
Robert Haberle, National Aeronautics and Space Administration
Lee Hunt, National Research Council
Jerry Mahlman, Geophysical Fluid Dynamics Laboratory
Robert Malone, Los Alamos National Laboratory
Michael MacCracken, Lawrence Livermore National Laboratory
Elizabeth Panos, National Research Council
Joyce Penner, Lawrence Livermore National Laboratory
James Pollack, National Aeronautics and Space Administration
Roseanne Price, National Research Council
S.J. Pyne, University of Iowa
Lawrence Radke, University of Washington
V. Ramanathan, National Center for Atmospheric Research
William Rose, Michigan Technological Institute
Stephen Schneider, National Center for Atmospheric Research
Donald Shapero, National Research Council
Renée St. Pierre, National Research Council
Starley Thompson, National Center for Atmospheric Research
Stephen Warren, University of Washington
Barbara Yoon, R&D Associates

Contents

1 Summary and Conclusions

The Committee on the Atmospheric Effects of Nuclear Explosions addressed the following charge: (1) determine the manner in which the atmosphere of the earth would be modified by a major exchange of nuclear weapons and, insofar as the current state of knowledge and understanding permits, give a quantitative description of the more important of the changes, and (2) recommend research and exploratory work appropriate to a better understanding of the question.

The committee was not asked to (and did not) address the related but distinct questions of the extent of radioactive fallout or the biological or social implications of postwar atmospheric modification.

Recent calculations by different investigators suggest that the climatic effects from a major nuclear exchange could be large in scale. Although there are enormous uncertainties involved in the calculations, the committee believes that long-term climatic effects with severe implications for the biosphere could occur, and these effects should be included in any analysis of the consequences of nuclear war. However, the committee cannot subscribe with confidence to any specific quantitative conclusions drawn from calculations based on current scientific knowledge. The estimates are necessarily rough and can only be used as a general indication of the seriousness of what might occur.

Despite the early state of understanding of these matters, the possibility of severe degradation of the atmosphere after a major nuclear exchange is of sufficient national and international concern that a major effort to narrow the scientific uncertainties should be given a high priority.

BACKGROUND

It is widely understood that any major nuclear exchange would be accompanied by an enormous number of immediate fatalities; nevertheless, a much larger fraction of the human population would survive the immediate effects of a nuclear exchange. This study addresses current knowledge about the nature of the physical environment the survivors would have to face.

The realization that a nuclear exchange would be accompanied by the deposition into the atmosphere of large amounts of particulate matter is not new. However, the suggestion that the associated attenuation of sunlight might be so extensive as to cause severe drops in surface air temperatures and other major climatic effects in areas that are far removed from target zones is of rather recent origin. That perception has grown out of a number of recent investigations. Crutzen and Birks (1982) recognized that the amount of smoke from the fires ignited by nuclear blasts could be of crucial importance, and Alvarez et al. (1980) hypothesized that the massive species extinctions of 65 million years ago were part of the aftermath of the lofting of massive quantities of particulates resulting from the collision of a large meteor with the earth. Others have recognized the similarity between the Alvarez dust hypothesis and the effects of nuclear war (see Appendix).

The consequences of any such changes in atmospheric state would have to be added to the already sobering list of relatively well-understood consequences of nuclear war, including prompt radiation, blast, and thermal effects, short-term regional radioactive fallout, inadequate medical attention for surviving casualties, and the long-term biological effects of global fallout. Long-term atmospheric consequences imply additional problems that are not easily mitigated by prior preparedness and that are not in harmony with any notion of rapid postwar restoration of social structure. They also create an entirely new threat to populations far removed from target areas, and suggest the possibility of additional major risks for any nation that itself initiates use of nuclear weapons, even if nuclear retaliation should somehow be limited.

THE COMMITTEE'S BASELINE CASE

To provide a framework for its study, the committee first constructed a baseline war scenario, made up of assumptions concerning the nature of the weapon exchange. The baseline scenario (see Chapter 3 for greater detail) was selected so as to be representative of a general nuclear war: one-half--about 6500 megatons (Mt)--of the estimated total world arsenal would be detonated. Of this, 1500 Mt would be detonated at ground level. Of the other 5000 Mt that would be detonated at altitudes chosen so as to maximize blast damage to structures, 1500 Mt would be directed at military, economic, and political targets that coincidentally lie in or near about 1000 of the largest urban areas. All explosions would occur between 30°N and 70°N latitude.

The committee also chose, on the basis of a review of the scientific literature, a set of baseline physical parameters to use in calculating the effects of the baseline weapon exchange. Each baseline parameter was chosen to lie well within the spectrum of scientifically plausible values, values in the middle ranges of plausibility being preferred.

There are three immediate consequences of a major nuclear exchange that could have a significant impact on the subsequent state of the atmosphere. Large amounts of dust could be lofted high into the atmosphere; large fires could be initiated; and large amounts of undesirable chemical species could be released. Some of the key parameters assumed for the baseline case follow.

The amount of dust (Chapter 4) that would be deposited in the stratosphere is related to total megatonnage. The committee assumes that the total amount lofted is 0.3 teragrams (1 Tg = 10^{12} g ≃ 10^6 metric tons) per megaton detonated. Eight percent of the mass of dust would be of submicron size, which remains aloft for long periods. The 1500 Mt in ground bursts would raise about 15 Tg of submicron dust into the stratosphere, where it could reside for more than a year. During that time, the solar radiation through that dust, and into the lower atmosphere, would be reduced.

The analysis of fires and smoke is complex (Chapter 5). The 5000 Mt of air bursts would initiate vigorous fires in cities and forests over areas where the thermal radiation incident on combustible material was 20 calories per square centimeter (cal/cm^2) or greater, a number well in excess of that known to be adequate to ignite the fuels at hand. In the city-scale urban conflagrations that would ensue, the baseline assumption is that three-quarters of the combustible material in affected areas would be consumed. (Nearly complete consumption of combustible materials is typical of large city-wide fires for which data are available.) Although many of the urban fires would probably spread beyond the 20 cal/cm^2 ignition zone, no additional fuel burden from that spreading is assumed in the baseline case. Of the material that burned in cities, the baseline case assumes that some 4 percent (limited data suggest values lying between 1 percent and 6 percent) would be converted to smoke particles in a range of submicron sizes that would absorb and scatter sunlight very effectively.

Certain processes may, however, diminish the optical effects of the smoke at this stage. During the burning of the urban and/or forest fires, the very fine smoke particles would undergo some coagulation in the rising plume. Over regions where the ambient ground-level humidity was high, the condensation of moisture entrained in the plume could incorporate some of the smoke. There is little empirical evidence to suggest extensive scavenging of the smoke by these processes, but in the baseline case, 50 percent of the smoke is assumed to be removed from the plumes of urban fires in this manner.

On the basis of available information on plume dynamics, it is assumed that, shortly after deposition the smoke from the ensemble of fires would be uniformly distributed vertically (mass per unit height) between 0 and 9 km over the entire affected area; the local vertical distribution would be nonuniform, however, because the altitudes of the smoke plumes would vary from one fire to another and would also vary with the time-dependent intensity of the fire. Although under special meteorological circumstances some of the smoke might be deposited at altitudes significantly higher than 9 km, this effect is ignored in the baseline case. Initially, and for some weeks, the smoke would have a very nonuniform horizontal distribution, but would be distributed throughout the troposphere of the northern temperate zone.

The injection of nitrogen oxides from the nuclear clouds into the upper atmosphere would lead to a depletion of the ozone column, which would be restored in about 2 years (Chapter 6).

The atmospheric implications of the baseline case (and of the results of other groups' analyses) are presented in Chapter 7. The committee expects that solar radiation passing through the stratospheric layer of nuclear dust would be absorbed in the upper regions of the smoke layer. The smoke layer would heat up, and since little solar radiation would reach lower levels, the air over land surfaces would cool.

Although a few types of natural events can provide marginally relevant information on aspects of the problem (see also Chapter 8), much of our understanding of the atmospheric response to large amounts of airborne particulates will come from model simulations. These models are validated within relatively small natural variations, so their predictive capability is limited for these large perturbations. Only preliminary estimates can be made of the rate of spreading of particulates over initially clear latitudes and of the rate of removal of particulates.

The duration and magnitude of atmospheric effects would depend on how long the absorbing particulates remained aloft. There is especially large uncertainty associated with long-term removal processes for smoke that survives the early scavenging. Low-altitude precipitation processes might remove the low-altitude smoke, that below 4 km (the normal range for smoke), rather efficiently. But at high altitudes, the increased air temperature and the low humidity could lead to a removal rate in the 4- to 9-km range that would be slower than the removal rate in today's troposphere. The baseline assumes that removal rates would be at least comparable to normal removal rates in the lower atmosphere (below 5 km), but would be slightly slower than normal in the upper troposphere (5 to 10 km), with about one-half of the initial particulates removed from the lower atmosphere in 3 days, and from the upper atmosphere in 30 days.* It is unlikely that the average residence times for postwar smoke would be much less than these values, and quite possible that the mean residence time in the upper troposphere would be longer.

It is hoped that the committee's baseline case will provide a useful point of departure for those who wish to identify and assess the environment that would prevail following a major nuclear exchange.

---

*If the smoke particles acquired electrical charges, the coagulation and smoke removal times could be affected. However, there is no evidence from large historical fires that electrical activity intense enough to be observed was operative. Furthermore, there is no evidence that the sometimes large and visible electrical effects in intense natural events (e.g., tornadoes and volcanoes) influence the dynamics of the storm. Thus, having no reliable basis on which to do otherwise, the committee has disregarded potential electrical effects.

NOTES ON THE NATURE AND SIGNIFICANCE OF UNCERTAINTY

As may be clear from this brief description of the baseline case, there are many points in the analysis at which there is a wide range of parameter values that are consistent with the best current scientific knowledge. Any estimate of the overall atmospheric response will involve a compounding of the effects of these uncertainties. Obviously, calculations made under these conditions cannot be read as a scientific prediction of the effects of a nuclear exchange; rather, they represent an interim estimate from which the reader can infer something of the potential seriousness of the atmospheric degradation that might occur.

Some reviewers of earlier drafts of this report cautioned that even the most qualified numerical results produced under these conditions could be misinterpreted, and some suggested that at present the only scientifically valid conclusion would be that it is not at this time possible to calculate the atmospheric effects of nuclear war. The committee believes, however, that an appropriately qualified, preliminary quantitative treatment of the problem is warranted on two grounds. First, given the enormous human stakes that may be involved, it may not be advisable to wait until a strong scientific case has been assembled before presenting tentative results; there is a danger that a report that reached no conclusions at all would be misconstrued to be a refutation of the scientific basis for the suggestion that severe atmospheric effects are possible. Second, a quantitative approach to the problem is the best way to ensure that all important factors are systematically considered, and quantification helps distinguish the important factors from the less important ones in the overall analysis. Such results are necessary to the orderly allocation of resources to the most pertinent research questions.

The findings of this report depend in rather large measure on a still limited body of scientific inquiry, some of which is not yet fully documented. Attention to the subject is so recent, in fact, that some of the underlying analysis has not yet undergone the peer review process that precedes publication in most scientific journals.

The reader should appreciate the possibility that further research may well invalidate some of the estimates discussed in this report. As recently as 1975, when the National Research Council report _Long-Term Worldwide Effects of Multiple Nuclear Weapons Detonations_ appeared, plausible weapon use scenarios differed significantly from those envisaged today, and the crucial importance of fires and smoke had not then been recognized. It follows that the findings presented in this report differ from those of the 1975 report. Furthermore, the pervasive uncertainties in the data and the limited validity of the atmospheric models used to date imply that some future study, conducted at a time when the data and models have been improved, could produce quite different analyses and conclusions. It is possible that improved understanding of some mechanisms (e.g., early scavenging) could so affect the results that the atmospheric degradation would be shown to be weaker than that estimated in the baseline case, but the same uncertainty also makes it a clear possibility that the exchange could produce a degradation that would be greater than, and would last longer than, that estimated in the baseline case.

In short: the committee's findings are clearly and emphatically of an interim character.

A vigorous research effort is now needed. Nevertheless, one cannot expect that long-term nuclear effects will be characterized with great precision or confidence in the next few years. Many uncertainties cannot be narrowed because they depend on human decisions that can be made, or changed, long after any particular prediction has been issued. These include, for example, the total yield of the exchange, individual warhead yields, the mix of targets, the mix of altitudes at which the bursts would occur, and the season of the year in which the exchange would occur. In addition, there are obvious limits to the use of large-scale experiments in this field, and the evolution of atmospheric models will require some time.

Many significant uncertainties, however, can be narrowed by further study. In particular, the heights to which smoke is deposited in city-scale fires, the early smoke removal by coagulation and condensation in the fire plume, the extent of continued buoyant rising of sun-heated opaque clouds, and the dynamical response of the atmosphere, first to patchy high-altitude solar absorption and then to the heating of more broadly distributed but still heavy smoke cover, have received only scattered and recent attention.

CONCLUSIONS

The general conclusion that the committee draws from this study is the following: a major nuclear exchange would insert significant amounts of smoke, fine dust, and undesirable chemical species into the atmosphere. These depositions could result in dramatic perturbations of the atmosphere lasting over a period of at least a few weeks. Estimation of the amounts, the vertical distributions, and the subsequent fates of these materials involves large uncertainties. Furthermore, accurate detailed accounts of the response of the atmosphere, the redistribution and removal of the depositions, and the duration of a greatly degraded environment lie beyond the present state of knowledge.

Nevertheless, the committee finds that, unless one or more of the effects lie near the less severe end of their uncertainty ranges, or unless some mitigating effect has been overlooked, there is a clear possibility that great portions of the land areas of the northern temperate zone (and, perhaps, a larger segment of the planet) could be severely affected. Possible impacts include major temperature reductions (particularly for an exchange that occurs in the summer) lasting for weeks, with subnormal temperatures persisting for months. The impact of these temperature reductions and associated meteorological changes on the surviving population, and on the biosphere that supports the survivors, could be severe, and deserves careful independent study.

A more definitive statement can be made only when many of the uncertainties have been narrowed, when the smaller scale phenomena are better understood, and when atmospheric response models have been constructed and have acquired credibility for the parameter ranges of this phenomenology.

The committee also draws several more specific conclusions:

1. In an extensive nuclear exchange, explosions over urban areas and forests would ignite many large fires. Massive smoke emissions are an important aspect of nuclear warfare that have only recently been recognized. For the major 6500-Mt nuclear war considered here, fires could release massive amounts of smoke into the troposphere over a period of a few days. Much of the smoke might be removed by meteorological processes within several weeks, depending on feedback effects, but significant amounts could remain for several months. During its tenure in the atmosphere, the smoke would gradually spread and become more uniformly distributed over much of the northern hemisphere, although some patchiness would be likely to persist. Light levels could be reduced to one percent in regions that were covered with the initial hemispheric average smoke load, causing intense cooling beneath the particulate layer and unusually intense heating of the upper layer. While large uncertainties currently attend the estimates of smoke emissions, and of their optical and physical consequences, the baseline case implies severe atmospheric consequences.

2. The production of smoke from fires, and the implied effects on the atmosphere, is more directly linked to the extent of detonation over urban areas than to the aggregate yield of a nuclear exchange. The industrialized nations of the world have concentrated a large proportion of their resources and combustible fuels in the vicinity of the central areas of their large cities. Any war scenario that subjects these city centers to nuclear attack, even one employing a very small fraction of the existing nuclear arsenal, could generate nearly as much smoke as in the 6500-Mt baseline war scenario.

3. The climatic impact of soot is very sensitive to its lifetime in the perturbed atmosphere and the uniformity of its distribution. The lifetime of soot is highly uncertain, particularly in the upper troposphere. The perturbation itself would produce severe new effects, many of which could tend to increase the residence time of the soot. Although the lofted soot (and dust) would rapidly spread around the latitude band of injection, the distribution could be uneven for several months, with continent-size patches of lesser and greater density, particularly near the southern edge of the affected zones.

4. In the baseline nuclear war scenario, hundreds of teragrams of dust would be injected into the atmosphere from surface detonations. A significant fraction of the dust consisting of particles with radii less than one micron (1 μm) would be expected to remain aloft for months. About one-half of these submicron particles would be injected into the stratosphere and would produce some long-term reduction of sunlight at the earth's surface, even after smoke and dust at lower altitudes were removed. This stratospheric dust alone would lead to

perceptible reductions in average light intensities, and continental surface temperatures would fall measurably. In a plausible scenario that involves more ground burst attacks against very hard targets than are assumed in the baseline case, the possible dust effects are several times larger.

5. It is not possible at this time to estimate the most probable average temperature changes at the surface caused by smoke and dust lofted in the baseline case; nor would such a single value, even if available, meaningfully describe the situation. In addition to the large uncertainties in many of the critical physical parameters and the inherent limitations of the models available for computer simulations, the available calculations reflect wide seasonal and geographical differences. Recent general circulation model simulations that incorporate simplifying assumptions indicate that a baseline attack during the summer might decrease mean continental temperatures in the northern temperate zone by as much as 10° to 25°C, with temperatures along the coasts of the continents decreasing by much smaller amounts. In contrast, an attack of the same size during the winter, according to these simulations, might produce little change in temperature in the northern temperate zone, although there could be a significant drop in temperatures at more southern latitudes.

6. The nitrogen oxides deposited in the stratosphere by nuclear detonations would reduce the abundance of ozone. For the 6500-Mt nuclear war, the northern hemisphere ozone reduction could become substantial several months after the war. Estimates based on current stratospheric structure suggest that the amount of ozone reduction would decrease by one-half after about 2 years. At the time of maximum ozone reduction, the biologically effective ultraviolet intensity (using the DNA action spectrum) at the ground would be approximately one and one-half times the normal levels. Initially, the presence of dust and smoke particles in the atmosphere would provide a measure of protection at the surface from the enhanced ultraviolet radiation. This protection would gradually diminish as the particles were removed.

7. This study has concentrated on the possible effects that a nuclear war could have on the northern hemisphere, primarily within the mid-latitude region (30°N to 70°N) where the nuclear exchange would be concentrated. It is particularly difficult to assess the potential effects of the baseline war on the atmosphere of the northern tropics and southern hemisphere. Although southern hemisphere effects would be much less extensive, significant amounts of dust and smoke could drift to and across the equator as early as a few weeks after a nuclear exchange. A large rate of transport across the equator driven by heating in the debris cloud cannot be ruled out. Indeed, such heating-enhanced cross-equatorial circulation has been found for spring and summer months in computer simulations.

8. Some prehistoric volcanic eruptions and impacts from extraterrestrial bodies have released energies corresponding to levels that would be released in a major nuclear exchange and may have lofted massive amounts of dust; however, neither type of event provides a useful direct analog to the nuclear case because neither type involved the production of highly absorbing soot particles. Furthermore, the

atmospheric consequences of prehistoric natural events of these proportions are not known, and their effects on the fossil record, if any, have not been sought in any systematic way. Accordingly, available knowledge about prehistoric volcanic and impact events provides neither support nor refutation of the committee's conclusions.

9. All calculations of the atmospheric effects of a major nuclear war require quantitative assumptions about uncertain physical parameters. In many areas, wide ranges of values are scientifically credible, and the overall results depend materially on the values chosen. Some of these uncertainties may be reduced by further empirical or theoretical research, but others will be difficult to reduce. The larger uncertainties include the following: (a) the quantity and absorption properties of the smoke produced in very large fires; (b) the initial distribution in altitude of smoke produced in large fires; (c) the mechanisms and rate of early scavenging of smoke from fire plumes, and aging of the smoke in the first few days; (d) the induced rate of vertical and horizontal transport of smoke and dust in the upper troposphere and stratosphere; (e) the resulting perturbations in atmospheric processes such as cloud formation, precipitation, storminess, and wind patterns; and (f) the adequacy of current and projected atmospheric response models to reliably predict changes that are caused by a massive, high-altitude, and irregularly distributed injection of particulate matter. The atmospheric effects of a nuclear exchange depend on all of the foregoing physical processes ((a) through (e)), and their ultimate calculation is further subject to the uncertainties inherent in (f).

REFERENCES

Alvarez, L.W., W. Alvarez, F. Asaro, and H.W. Michael (1980) Extraterrestrial cause for the Cretaceous-Tertiary extinction. Science 208:1095-1108.

Crutzen, P.J., and J.W. Birks (1982) The atmosphere after a nuclear war: Twilight at noon. Ambio 11:114-125.

National Research Council (1975) Long-Term Worldwide Effects of Multiple Nuclear Weapons Detonations. Washington, D.C.: National Academy of Sciences.

2 Recommendations for Research

The uncertainties that impede an accurate quantification of the atmospheric degradation implied by a nuclear exchange are numerous and are large. Nevertheless, despite the limited understanding of these matters, it is clear that the hazard posed by a large nuclear exchange is of serious concern and deserves continuing attention.

Accordingly, investigations should be initiated to narrow the uncertainties and to provide credible, quantitative, and reasonably accurate estimates of the atmospheric modifications. To that end the committee makes the recommendations below. These recommendations are not intended to define precisely designed studies, but rather to indicate the general areas in which it appears that the uncertainties can most probably be narrowed. Because further uncertainties and even new causal mechanisms could emerge, research programs that address this problem should be flexible enough to accommodate changing scientific information needs.

1. Our ability to predict the climatic consequences of prescribed injections of smoke and dust into the atmosphere is not well developed at this time, but it can advance substantially in the next few years. Problems that deserve special emphasis are (a) the transport of smoke and dust by the atmosphere and the subsequent feedback effects of this transport on the circulation, (b) the lifetime of smoke particles in the perturbed atmosphere, (c) the transient response of the surface temperature and near-surface meteorology to fluctuating light levels on day to week time scales, (d) the regional influence of ocean-continent climate interaction, (e) the optical and infrared properties of smoke aerosols and their evolution with time, and (f) the implications of changes in precipitation and cloud structure.

One-dimensional radiative-convective climate models have proven very useful in the initial phase of study, but, except for problem (c), are severely limited in their ability to address these problems. Two-dimensional zonally symmetric circulation models can be used to study areas (a) and (b), but the limitations imposed by not easily validated parameterizations of large-scale atmospheric eddies must be kept in mind. Three-dimensional general circulation models (GCMs) could be used to address all of the problems above, although existing models are not properly designed for the task. The groups that have such models should be encouraged to use their resources in an effort to decrease the current level of climate modeling uncertainty. Additional supporting research will be needed to improve the treatment of those processes in GCMs that influence aerosol lifetime and surface conditions.

2. There is a need to develop a better understanding of the following phenomena related to the fires induced by nuclear war: (a) the dynamics of large-scale fires, including fire spread, the effects of prompt water condensation, and the implications for smoke injection heights, optical properties, and removal rates; (b) inventories of urban fuels by type and density; (c) seasonal variations in, and areal extents of, wildland fuel burdens (using existing surveys and Landsat data, for example); (d) the efficiency of ignition of both urban and natural fuel arrays by nuclear bursts; (e) the quantity, composition, and size distribution of smoke produced by different fuels in large-scale fires, and the extent of fine particulates other than smoke swept into fire plumes; (f) the microphysical processes that remove smoke particles from the atmosphere in the long term; and (g) the mesoscale transport and mixing of smoke, including the coupling of smoke heating and microphysics to dynamics and precipitation.

These recommendations concerning fires and smoke are only generally spelled out, as it was not the objective of the committee to design specific research programs. However, it is worth emphasizing that a particularly important subject that is amenable to both experimental and theoretical study is the microphysical processing (by coagulation and water scavenging) of smoke particles in large fire plumes, and the implication for radiative properties of the smoke layer. Because of the complexity of the fire/smoke problem, research planning and management in this area should be carefully coordinated.

3. An important effect of nuclear war is radioactive fallout. This topic does not lie within the scope of the present study, and the committee has not addressed it; but the committee does believe that the radioactivity problem should be reexamined in appropriate detail. Earlier studies estimated that the globally averaged long-term radiation hazard would not be large. However, these studies depended on an analysis of strontium-90 fallout from atmospheric nuclear tests, which were dominated by multimegaton bursts that resulted in stratospheric injection of the radioactive debris, and thus avoided early rainout. Current arsenals comprise mainly submegaton warheads, which would inject a major share of their radioactivity into the mid-latitude troposphere. Thus widespread fallout and precipitation on intermediate time scales (days to weeks) might pose a more serious problem. New preliminary estimates suggest that low-yield scenarios significantly increase the globally averaged radiation hazard per megaton released (Knox, 1983; Turco et al., 1983).

4. Although the committee's conclusions on dust seem well constrained, they are largely based on a small body of empirical data derived from nuclear tests that in some respects are imperfect analogs of 0.5- to 1.0-Mt continental surface bursts. The data come from either high-yield tests at the Pacific Proving Ground or from Johnie Boy, a very low yield, partially buried test in Nevada. More detailed

theoretical treatments (especially, detailed numerical simulations) of fireball mass loading (i.e., vaporized and melted rock and ejecta), lofting, and recondensation processes that lead to the final distribution of dust particles in the stabilized cloud could put these estimates on a firmer footing.

5. With the exception of one study using a two-dimensional model, all of the published calculations of ozone decreases from nuclear explosions (including the present set) have come from one-dimensional models of chemistry and transport. Present-day scenarios of nuclear war involve the injection of nitrogen oxides in the lower stratosphere, where the chemical processes and the transport processes both have similar characteristic times. Suitable two- and three-dimensional models potentially can represent the ozone space-time variations more realistically than a one-dimensional model. The committee recommends the development and application of higher-dimensional models to this problem. Included in the work should be appropriate sensitivity tests and a degree of verification of the model.

6. Geologists and paleoclimatologists should make joint investigations on the local and global scales of the possible climatic effects of well-dated great volcanic explosions that have occurred in the past few million years. The errors in dating volcanic eruption products increase with increasing age: hence the youngest great volcanic explosions offer the best opportunity to correlate climate changes with particular explosions. Research is also needed to quantify adequately the character and amount of material injected into the atmosphere by large volcanic explosions.

Recent work suggests that the occurrence of regional to global effects of meteorite impacts in the energy range 10^6 to 10^7 Mt may be recognizable in marine and continental sediments. Multidisciplinary investigations of identifiable events should be carried out to assess the effects of large impacts on climate and on the biology.

REFERENCES

Knox, J.B. (1983) Global scale deposition of radioactivity from a large scale exchange. Paper presented at the International Seminar on Nuclear War, 3rd Session: The Technical Basis for Peace. Ettore Majorana Centre for Scientific Culture, Erice, Sicily, August 19-24, 1983.

Turco, R.P., O.B. Toon, T.P. Ackerman, J.B. Pollack, and C. Sagan (1983) Global Atmospheric Consequences of Nuclear War. Interim Report. Marina del Rey, Calif.: R&D Associates. 144 pp.

3
The Baseline Nuclear Exchange

The conclusions of any study of the consequences of nuclear war depend on the level and nature of the weapons exchange. The baseline case for this study, consistent with the mission statement, depicts a major nuclear war between the United States and the Soviet Union. The committee has not chosen the baseline assumptions to depict either the "most likely" general war scenario or the "worst-case" general war scenario. In defining the baseline case, the committee has sought to establish a credible, generalized account of the extent of a possible general nuclear war in the mid-1980s; hence it is not necessary to specify the manner in which this general war might begin or might escalate from the initial use of nuclear weapons or to designate specific weapons for specific targets.

United States and Soviet nuclear forces reportedly now include about 50,000 nuclear weapons, with a total yield of some 13,000 Mt. About 25,000 of these nuclear weapons, with a yield of about 12,000 Mt, are on systems with strategic or major theater missions. The other 25,000 weapons, mostly of much smaller yield, are designed for tactical battlefield, air defense, antisubmarine, naval, and other special missions. In this analysis the committee has assumed (see Table 3.1) that approximately one-half of these weapons, or 25,000, would actually be detonated, with a total yield of about 6500 Mt. This would include 12,500 strategic and major theater weapons with a yield of 6000 Mt and 12,500 tactical weapons with a yield of 500 Mt. The fraction of one-half has been applied to take into account the following factors that would reduce the number of weapons actually delivered on target: weapons destroyed by counterforce attacks, weapons destroyed by defenses, weapon systems unreliable under combat conditions, and weapons held in reserve. This assumption should be within a factor of 2 of the exchange in a general nuclear war.

The weapons in this exchange are all assumed to be 1.5 Mt or less, with a major fraction less than 1.0 Mt. This represents a shift from many earlier analyses, which included significant numbers of 10- and 20-Mt bombs and missile warheads. The elimination of very high yield weapons reflects the fact that both nations have, in recent years, been increasing the accuracy and fractionating the payloads of their missiles to obtain larger numbers of lower yield warheads. Similarly, multimegaton bombs have been replaced by more and smaller bombs and by

TABLE 3.1 Baseline Case: Weapon Size Distribution

Yield per Warhead (Mt)	Number	Total Yield (Mt)
A. Total Detonated		
1.5	220	330
1.0	2,700	2,700
0.75	330	250
0.5	3,600	1,800
0.25	2,400	600
0.05 to 0.15	3,200	320
Tactical (misc.)	12,500	500
	~25,000	6,500
B. Ground Bursts		
1.0 to 1.5	400	500
0.5	2,000	1,000
	2,400	1,500

large numbers of stand-off cruise missiles with smaller yields. By 1985, there will probably be few, if any, multimegaton weapons deployed by either the United States or the Soviet Union, unless present trends are reversed.

In a general nuclear war between the United States and the Soviet Union, the committee has assumed that all member nations of NATO and the Warsaw Pact would be involved and targeted for strategic weapons. The significance of this assumption to the study is that a number of targets located in urban areas, which are the major source of smoke, are found outside the United States and Soviet Union. It is further assumed that tactical nuclear war would for the most part be confined to the NATO/Warsaw Pact area (European Front) and the oceans. While other key allies and countries could well become involved in such a conflict, the committee did not have a specific military rationale for including targets in these nations. Moreover, modest numbers of military targets in such countries would not significantly alter the study results.

The description of specific targets in all of these countries for 12,500 strategic and major theater weapons would be a difficult undertaking with no enduring validity. Even if the specific targeting plans of the nuclear powers were adopted, such detail could be misleading in suggesting that there would be a unique predictable pattern to a general nuclear exchange. Moreover, such detail is not relevant to this study, which relies on models that do not have as inputs the actual locations of targets. Factors such as proximity to oceans might be important to more sophisticated future models.

The committee has assumed that each side would give highest priority to "counterforce" attacks against the vulnerable components of the other side's threatening strategic forces and against the command, control, communications, and intelligence (C^3I) facilities necessary to operate those forces effectively. It is also assumed that high priority would be given to destroying key military bases and transportation and communications nodes necessary for theater operations, particularly in Europe. The committee has assigned approximately 9000 effective warheads with a yield of some 5000 Mt to these missions. This would be consistent with each side's attacking each of the other side's strategic missile silos with two weapons in order to improve the kill probability; multiple attacks on several hundred military and civilian airfields capable of sustaining redeployed strategic aircraft; multiple attacks on submarine and naval bases; extensive attacks against the central civilian and military command and control systems, the critical nodes in the military communications system and facilities necessary to exploit intelligence assets for real-time targeting and damage assessment; and multiple attacks on several hundred major theater military targets.

The committee has assumed that each side would, as a second priority, attack the other's economic base necessary to sustain its military efforts. These "countervalue" targets would include plants producing military equipment, important components, and materials, petroleum refineries and storage, and electric power plants, as well as key transportation and communication nodes. In this scenario, some 3500 effective warheads with a yield of 1500 Mt would be used against such targets.

While neither side would target population per se, the committee has assumed that neither would refrain from attacking urban areas if military or economic targets were located there. Most economic targets are co-located with urban areas, and many military targets, such as airfields capable of sustaining redeployed strategic aircraft, naval bases, and C^3I facilities, are also co-located with urban areas. The number of economic targets not co-located with urban areas may be comparable to the number of military targets that are co-located with urban areas. Therefore, for the purpose of this study the committee has assumed that some 3500 weapons with a yield of approximately 1500 Mt would strike urban areas. Specifically, as a first approximation, it is assumed that economic targets and co-located military targets would be distributed in the largest 1000 NATO/Warsaw Pact urban areas roughly in proportion to the population of those areas. As detailed in the chapter on fires resulting from such an attack, it is assumed that there would be one-third overlap of areas exposed to 20 cal/cm^2. These assumptions imply that fire ignition would occur over 50 percent of the areas of these cities.

The committee has assumed that both sides would fuze their warheads for air or ground burst to optimize military effectiveness against the targets under attack and not to increase population fatalities. With this in mind, it is estimated that about 25 percent (1500 Mt) of the total yield would be ground bursts. One ground burst is assumed against each silo and other hardened target.

Given the large number and wide distribution of possible targets in this scenario, it is assumed as a first approximation that the targets and megatonnage would be distributed evenly over the land areas from latitudes 30°N to 70°N. A more precise approximation of this distribution of megatonnage could be determined by examining the density of known major strategic targets and urban areas within these latitudes; however, such detail would not add appreciable precision to the present estimation of atmospheric consequences until knowledge about soot production, transport, and removal is much improved.

It is important to note that this weapons exchange assumes that all targets would have been chosen to have direct or indirect impact on the ability of the two sides to conduct or sustain military operations or to emerge from the hostilities in a superior position. No targets would be chosen to maximize worldwide population fatalities or long-term effects on the biosphere. Consequently, it is assumed that there would be no attacks on urban areas in countries not directly involved in the conflict. The committee has assumed that there would be no attacks solely designed to ignite or sustain forest fires--and no attacks on oil fields, since the destruction of storage facilities and refineries would provide more immediate and effective denial of petroleum products. In addition, it is assumed that the war at sea would be directed against specific ships and submarines.

In this 6500-Mt baseline case, no large multimegaton weapons would be employed by either side. In order to examine the atmospheric effects of very high yield explosions, the committee has also analyzed a second case--an 8500-Mt excursion--in which sufficient multimegaton (i.e., 20 Mt) missile warheads would be deployed to permit successful delivery of approximately 100 such weapons on superhard, high-value targets, in addition to the 6500-Mt baseline megatonnage. It is assumed that these would all be surface bursts.

4
Dust

Part of the particulate burden of the atmosphere that would result from a nuclear war would be a collection of recondensed rock and metal vapor, small spheres of glass formed from quenched rock melt, and unmelted mineral and rock particles, all loosely termed "dust." Surface bursts are much more effective than air bursts or buried explosions in raising dust to great altitude; explosions in the megaton range would loft dust into the stratosphere, where its residence time could be many months. (A general discussion of the dynamics of atmospheric nuclear explosions can be found in Glasstone's _Effects of Nuclear Weapons_ (Glasstone and Dolan, 1977) and in Brode's (1968) review.) As will be detailed below, estimates of the dust that would be lofted by a high-yield surface burst lie in the range 0.2 to 0.5 teragrams (1 Tg = 10^{12} g ≃ 10^6 metric tons) per megaton (Mt) of explosion energy; a likely value would be 0.3 Tg/Mt. For a nuclear exchange including 1500 Mt of surface bursts, the total lofted mass could be 330 to 825 Tg. Roughly 8 percent of the dust would be in the submicron radius range. The submicron fraction could range from a few percent to 20 percent. A plausible value for the mass of lofted submicron dust would be 40 Tg.

The following sections of this chapter summarize the dynamics of nuclear clouds, the source mechanisms of dust, the available data that give estimates for dust lofting, and finally, an analysis of dust lofting in this baseline case.

NUCLEAR CLOUD DYNAMICS

Unlike soot from the long-lasting fires, dust from a nuclear explosion would be lofted to its stabilization altitude within 3 to 4 min, and, once a nuclear attack stopped, there would be no additional sources. Dust has an appreciable effect on climate only if it is of small size (submicron, or less than one micrometer (1 μm) in radius) and if it is lofted to the stratosphere, where residence times are appreciable. (An altered state of the atmosphere would make estimates of residence times less certain. Consideration of dust lofted to all altitudes is required in climate simulations.) Lofting into the stratosphere requires a substantial explosion energy, a yield above roughly 1 Mt.

Most of the following discussion will be based on idealized calculations for 1-Mt surface bursts (Zinn, 1973; Horak et al., 1982; Horak and Kodis, 1983).

The characteristics of nuclear explosions have been understood since the early years of the nuclear age, but detailed unclassified accounts have appeared only in recent years. Zinn (1973) has calculated the early growth of a 1-Mt fireball, assuming a hypothetical device with a mass of 1 metric ton. The radiation temperature of the exploding bomb was chosen to be 1.6 keV (1.9×10^7 K); the explosion produced a stream of X-rays with a peak in its blackbody distribution at about 4.5 keV (5.2×10^7 K). In air at room temperature, the mean-free path for such photons is a small fraction of a meter, so the first photons emitted heated the air around the bomb, causing the opacity of this layer to drop. Subsequent photons traveled freely across the heated layer until they reached the edge, where they, in turn, were absorbed. In this way the fireball grew until the emission of X-rays by the bomb stopped; Zinn assumed this happened at 0.1 microsecond (μs). The fireball was still very hot at the end of the X-ray deposition phase, with a temperature of about 1 keV (1.2×10^7 K), and was itself a source of high-energy photons. The fireball grew by radiative (photon) transport until the growth rate was no longer large in comparison with the sound speed in the heated gas. Until this time, any hydrodynamic signal that formed at the outer pressure discontinuity would be outrun by the radiative expansion. By 50 μs, however, the fireball had cooled sufficiently that a shock wave formed. The fireball radius at that time was about 50 m. Subsequent growth was dominated by hydrodynamic phenomena, even though the fireball eventually lost about 30 to 40 percent of the total yield in the form of optical photons. By the time the fireball radius had doubled to 100 m (at about 2 milliseconds (ms)), the expansion could be described by classical blast-wave theory (Bethe et al., 1947; Taylor, 1950; Brode, 1955; Sedov, 1959).

Fireball expansion reduced the internal pressure (radiative losses contributed a relatively minor reduction) until, at a few seconds, the pressure reached atmospheric levels. The shock wave continued to expand and weaken, but fireball growth stopped at a radius of about 1 km. For a 1-Mt surface burst, fireball growth would have stopped at about 1.3 km; the fireball would behave approximately as if its yield were 2 Mt, because the ground surface acts as a mirror (see Glasstone and Dolan, 1977, p. 71).

Because the density of the heated air in a fireball is low (about 4 percent of atmospheric density), the fireball is so buoyant that it would rise a distance equal to its own radius in the first 10 s. Deformation due to buoyancy and, in the case of a low air burst, to the effects of the shock wave reflected from the surface would transform the fireball into a ring vortex, which is the cap of a mushroom cloud. As the fireball rises, air near the ground would move inward (the "afterwind") and then up, creating the stem of a mushroom cloud. Details of fireball rise (typically at about 100 m/s) are determined by the density structure of the atmosphere, entrainment rates, wind shears, atmospheric humidity, and other factors (Sowle, 1975). However, the general character of the rise is a result of balance between the downward force of the drag interaction between the fireball and the atmosphere and the upward buoyant force as follows:

$$\frac{1}{2}\rho_a v^2 K\pi R^2 = \frac{4}{3}\pi R^3 \rho_a g, \tag{4.1}$$

where ρ_a is ambient atmospheric density, v is the rise rate, K is a drag coefficient (K ≃ 2; e.g., Öpik, 1958), R is a characteristic fireball radius, and g is the acceleration of gravity. The density in the fireball has been assumed to be negligible. This reduces to

$$v = (8Rg/3K)^{1/2}; \tag{4.2}$$

for a 1-Mt burst the radius is 1 km, giving a rise rate of 115 m/s, a value in good agreement with observations.

Because the density of the expanding vortex would not drop with altitude as fast as the density of the atmosphere, the fireball would eventually lose buoyancy and, after a few minutes, stop rising at the so-called stabilization altitude. Figure 4.1 (from Glasstone and Dolan, 1977, p. 32) illustrates the rise of a 1-Mt air burst, and Figure 4.2 shows the measured heights of cloud tops for a series of high-yield nuclear weapons near the equator in the Pacific. These heights are somewhat less than those quoted by Foley and Ruderman (1973) and used by Turco et al. (1983). The present data are derived from original sources (EG & G Technical Staff, 1958). Higher latitude bursts would show less rise, because of the different atmospheric structure (Figure 4.3). Stabilization heights for continental bursts would also be lower than the heights observed for the Pacific tests,

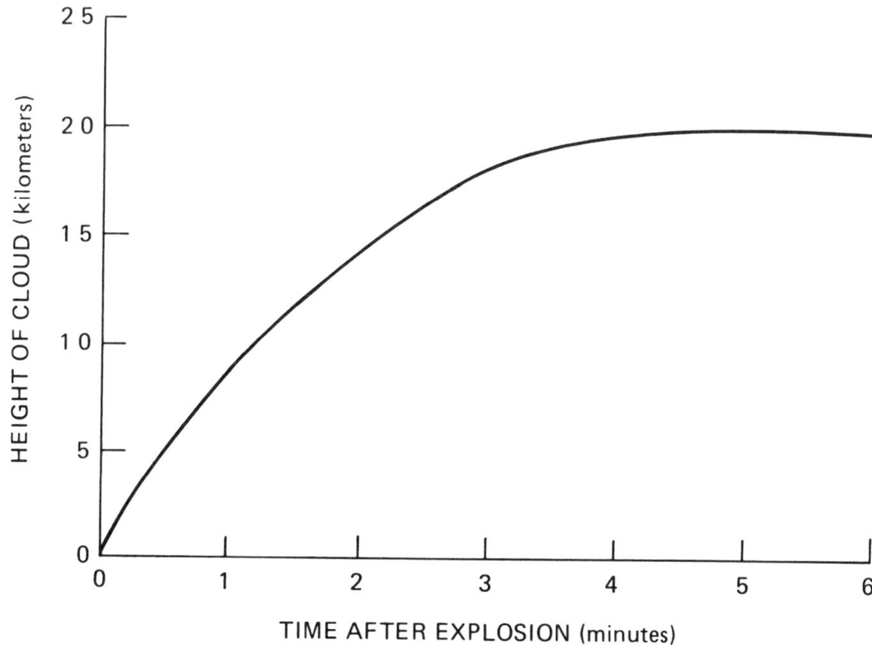

FIGURE 4.1 Height of the top of a cloud produced by a 1-Mt low-altitude burst (after Glasstone and Dolan, 1977).

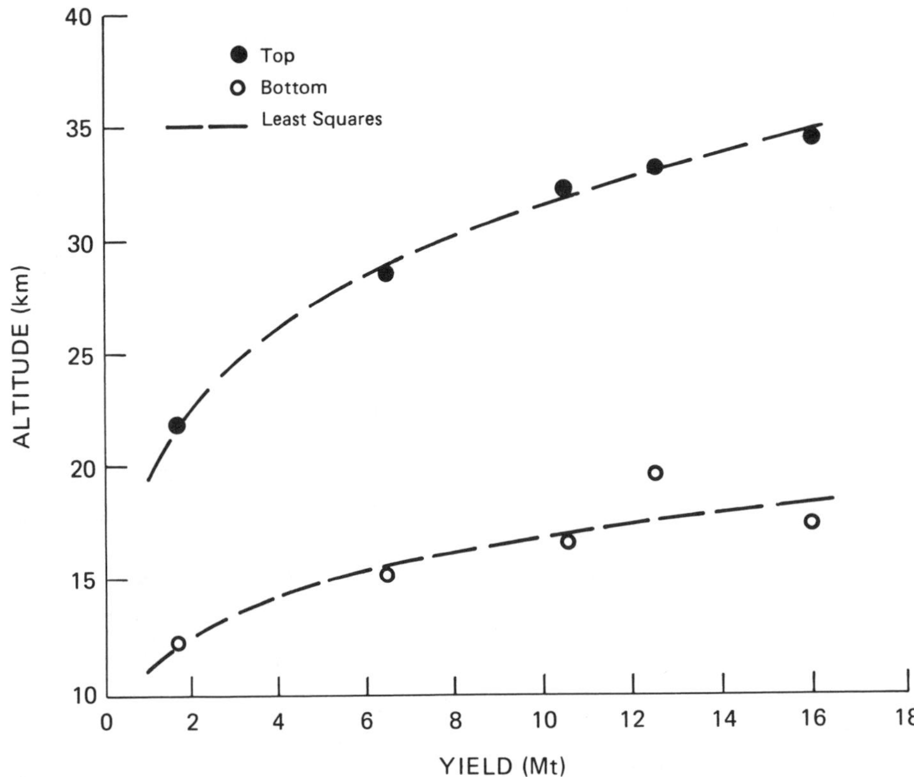

FIGURE 4.2 Heights of the cloud tops and bottoms at 10 min for the high-yield Pacific tests done during operation Castle. The dashed curves are least-squares fits to the data. Stabilization altitude is reached by these clouds in about 6 to 7 min, with most of the rise occurring during the first 3 min.

because of the smaller water content in fireballs over continents; water vapor condensation releases energy into the cloud, increasing buoyancy. If detailed descriptions of stabilization altitudes and stratospheric penetration unexpectedly prove to be of interest in numerical simulations, parameterization based on assumed yields, burst heights, and atmospheric structure could be derived from the model of Sowle (1975).

The preceding discussion of fireball rise applies only to explosions with total energies below about 150 Mt. Although no recent explosion, man-made or natural, has approached 150 Mt, conceivable multiburst attacks on very hard targets as well as some natural explosions such as meteor impacts could exceed this energy. Theoretical analysis suggests that at yields higher than 150 Mt, the resultant "giant" fireball would not be confined to the lower atmosphere and could rise at speeds of kilometers per second to hundreds of kilometers altitude, perhaps carrying great quantities of dust (Jones and Sandford, 1977; Jones and Kodis, 1982; Melosh, 1982; C.E. Needham, S-Cubed, Inc., Albuquerque, unpublished numerical simulations of 500-Mt explosions, 1982). The phenomenon may be

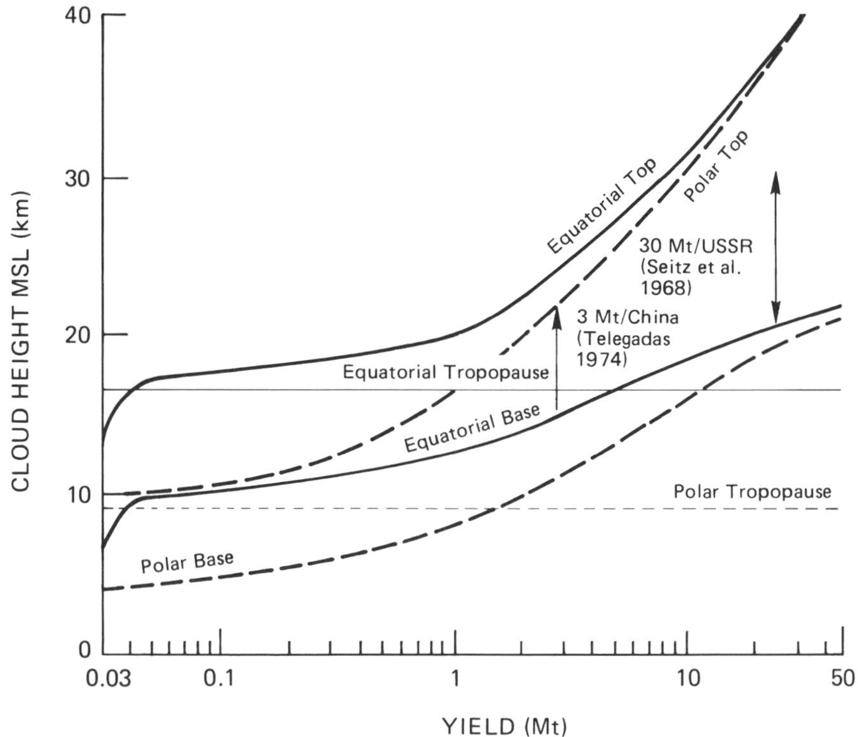

FIGURE 4.3 Estimates of differences in cloud rise between equatorial and high-latitude bursts (after Peterson, 1970).

understood as a ballistic rise occurring because the atmosphere is unable to confine the fireball explosion. Such a situation may account for the very large mass of dust lofted by the impact proposed by Alvarez et al. (1980, 1982) to explain the Cretaceous-Tertiary boundary claystone, and it might also arise, on a smaller scale, during concentrated nuclear attacks on hardened structures and missile silos. These situations will be discussed in more detail in the section on excursions below.

DUST LOFTING BY A NUCLEAR CLOUD

A rising fireball can carry gas and relatively large particles to great altitude. We have seen (Figure 4.1) that a 1-Mt burst would rise about 18 km in 3 min at an average rise rate of about 100 m/s. Such a flow could loft particles of up to a few centimeters in size.

An air burst fireball rises because the buoyant energy (a potential energy deficit in the earth's gravitational field) is converted into kinetic energy of the rising mass. The rate at which this conversion occurs is $\rho_a Vgv$, where V is the fireball volume. A surface burst would incorporate vaporized rock, molten rock, and solid particles in the low-density fireball. Lifting this mass requires the expenditure of some of the buoyancy energy. The fireball dynamics would be significantly altered (buoyancy would be eliminated) if the added mass of dust equaled the mass deficit of the fireball; i.e., M(dust) = $\rho_a V$. For a 1-Mt burst the radius of the low-density central region of the fireball would be about 800 m, which gives an upper limit of 2.6 Tg to the dust load that may be lifted by buoyancy. The actual loading would be about a factor of 10 less, a probable result of such factors as the limit of energy available for vaporization of rock, poor drag coupling of the large particles that dominate the mass distribution of crater ejecta, and particle-particle interactions.

Analysis of particle samples obtained by aircraft from the stabilized clouds of high-yield surface bursts detonated over water and/or coral islands indicates that the clouds loft about 0.2 Tg/Mt (Gutmacher et al., 1983). This value applies to yields of about 1 Mt or greater. There is a suggestion in the data, particularly for the lower yield test Koon (110 kt), that the lofting efficiency increases at lower yields. Koon lofted 0.5 Tg/Mt. The uncertainties in these measurements are probably rather large, but in the absence of additional information, particularly for high-yield continental surface bursts, a plausible range of lofting efficiencies is 0.2 to 0.5 Tg/Mt, with lower values favored for higher yields. A likely value for the 0.5- to 1.0-Mt surface bursts of the baseline scenario is 0.3 Tg/Mt. The lofting efficiency is not likely to approach the extreme upper limit of 2.6 Tg/Mt; at such levels the central density would approach ambient values, buoyancy would be reduced, and dynamics would be seriously perturbed. Fireball rise is observed to be consistent with a low value of dust loading.

SOURCES OF DUST

If a nuclear fireball is to raise significant amounts of dust to great altitude, the burst must occur very close to the ground. One measure of the ability of a fireball to raise particles is the amount of fallout observed near the explosion. This local fallout consists mostly of the largest particles, those that cannot be long supported by the flow and that fall to the ground early in the cloud rise. Glasstone and Dolan (1977) give an upper bound for a burst height giving significant local fallout at 870 $W^{0.4}$ m, where W is the yield in megatons. This is about one fireball radius. However, this does not mean that fireballs that fail to touch the ground produce no fallout and loft no small particles to stabilization altitude. It does mean that the lofted mass drops dramatically with increasing burst height. A practical limit is given by the observation that 10-kt bombs suspended at an altitude of 450 m beneath tethered balloons produced dusty stems that did not merge with the ring vortex (the mushroom cap). If this behavior scales with fireball radius and, hence, as $W^{0.4}$, a 1-Mt burst would not produce a stem/cap connection for a burst height above 3 km. Because this is near the detonation altitude of maximum blast damage, many bursts in a war would produce only dusty stems that would not connect with the fireball. Note that of the mechanisms described below, only "sweep-up" and thermal dust are likely to contribute to stem dust.

For bursts in the air, those very close to the ground ("surface bursts") are most effective in raising dust. If the weapon were slightly buried, the total mass in the cloud would increase dramatically, but because much of the explosion energy is deposited in the ground and there is no radiative fireball, the cloud rise would be very modest. A surface burst can be loosely defined as one close enough to the ground that the primary interaction with the soil occurs through the agency of radiative transport instead of blast. The details will depend on the radiative characteristics of the specific weapon, but from Zinn's (1973) hypothetical 1-Mt case it can be estimated that the burst height would have to be less than a few tens of meters.

X-rays would be deposited in a thin layer of rock or soil and would generate an intense shock wave in the ground. Close to ground zero, rock would be vaporized by the shock; farther out, rock would be melted; and finally, at greater distances, the rock would be displaced, creating a cloud of ejecta from the forming crater. All these processes would contribute to the dust load of the fireball. There are three additional sources of dust: the metal vapors that are the physical remains of the weapon, soil lofted in the so-called "thermal layer," and dust swept into the stem and fireball by afterwinds. These three mechanisms are not expected to be major sources of dust for surface bursts.

Recondensed vaporized material is an important source of fine particles in nuclear clouds from surface bursts. Most of the vapor is derived from rock and soil. Only a modest amount of metal is contained in a ballistic missile warhead.

The relative importance of the mechanisms that produce vapor from rock and soil varies with height of burst. If the bomb were exploded at or slightly below the surface, about half or more of the energy would be delivered as a strong shock propagated into the ground. Initially, this shock would be strong enough to vaporize rock. From calculations by Butkovich (1974) for underground explosions, the amount of vaporized rock produced by a surface burst may be estimated at 0.04 Tg/Mt for a dense rock target (density of 2.6 g/cm^3) and approximately 0.06 Tg/Mt for a porous dry soil or a very porous dry rock target (density of 1.4 g/cm^3).

In addition to vapor, a much larger mass of melted rock would be produced by the shock. For a surface burst on a dense rock target, about 0.5 to 0.6 Tg/Mt of rock would be shock melted; up to twice as much melt would be produced from porous targets. About half of the melt would be sprayed as a conical sheet out of the expanding crater. Both sides of the sheet would then be exposed to radiation from the fireball. Because temperatures in the early fireball would exceed the vaporization temperatures typical of rock melts (0.4 eV, or about 5000 K), part of the ejected melt sheet would be vaporized. In a 1-Mt explosion the temperature of the fireball would drop below typical vaporization temperatures for rock melts after about 5 s. Local fireball temperatures adjacent to the melt sheet would drop below vaporization temperatures sooner, owing to transfer of energy to rock vapor and to increased opacity near the melt sheet. The enthalpy required to vaporize silica melts is of the order of 500 calories per gram (cal/g), and, if all the energy of the fireball were transferred to the rock vapor, the entire melt sheet from a dense target would be vaporized (about 0.3 Tg/Mt). The temperature of the fireball would drop below the vaporization temperature of the melt sheet long before this could happen, however. The thin leading edge of the melt sheet, which would be exposed longest and to the highest energy radiation, probably would be entirely vaporized, but negligible vaporization would occur from the late, thick trailing part of the ejecta sheet. From rough considerations of the geometry and velocity structure of the ejecta sheet and the temperature history of the fireball, it is estimated that probably no more than about one-tenth of the melt sheet, (0.03 to 0.06 Tg/Mt) would be vaporized by radiation from the fireball.

The total amount of vaporized rock (shock-vaporized plus vaporized melt) expected from a surface burst therefore is of the order of 0.07 to 0.12 Tg/Mt, depending on the porosity and compressibility of the surface material.

The melt would also be the source of another class of small particles after the fireball cooled below the vaporization temperature. Divergent flow and aerodynamic disruption would break up the ejected melt sheet into droplets. Some of these droplets would remain sufficiently large that they would soon fall out of the fireball, but microscopic droplets would also be formed. The aerodynamic pressure, P_a, on the leading edge of the ejecta sheet is given approximately by $P_a = 1/2\ \rho_f v_e^2$, where ρ_f is the density of the fireball and v_e is the velocity of the leading edge. Initial ejection velocities are of the order of 10^5 cm/s, and the average density of the fireball, over time, is about 10^{-5} g/cm^3. Hence the initial aerodynamic pressure at the leading edge is of the order of 10^5 dyn/cm^2, sufficient to convert large droplets into a fine mist (acceleration of centimeter-size droplets is of the order of 10^5 cm/s^2). The physics of disruption of the ejected melt sheet is complex and is not understood in detail. It appears likely, however, that the mass of material carried up in the fireball as fine droplets or mist of shock-melted material is comparable to the mass of rock vaporized. This melt generally would be quenched to glass as the fireball cools. The total mass of small particles derived from the rock vapor and melt would be roughly 0.2 Tg/Mt, an estimate in good agreement with the observations described below.

The principal remaining sources of dust are solid particles ejected from the crater or swept up by the afterwinds. The size distribution of solid particles ejected from a surface burst crater is dependent on the characteristics of the target. Even from a crater produced in massive strong rock, a small fraction of the ejecta consists of micron and submicron particles. Ejecta from laboratory-scale craters in massive basalt consist about 1 percent by mass of particles finer than 10 μm; about 0.1 percent is finer than 1 μm (Gault et al., 1962; Moore and Gault, 1965). A detailed analysis of fragments produced by a 10-ton conventional explosion in massive tuff showed that 6 percent by mass of the particles was finer than 10 μm and 3 percent was finer than 1 μm (E. Shoemaker, USGS, unpublished manuscript, 1983). Most of the fine particles could be separated from the coarser fraction only by washing and ultrasonic agitation, however. About 0.5 percent of the total mass in particulates finer than 10 μm and 0.1 percent in particles finer than 1 μm were easily separated by simple mechanical agitation.

Most fine particles ejected from surface burst craters collide with and stick to larger fragments. As an upper bound, probably no more than about 1 percent of the total mass consisting of particles smaller than 1 μm is carried to stabilization altitude in the fireball from a surface burst on a strong rock target. The mass of material excavated from a crater produced by nuclear surface burst is sensitive to small differences in height of burst and to properties of the target, but is roughly of the order of 10 Tg/Mt (cf. Cooper, 1977). Hence the mass of the solid particles finer than 1 μm that are carried to stabilization altitude has an upper bound of the order of 0.1 Tg/Mt.

Ejecta from craters produced in fine particulate target material, such as fine alluvium, may be expected to yield somewhat more than 0.1 Tg/Mt of fine solids entrained in the fireball, provided that the target is dry. In ejecta from wet targets, on the other hand, the mass of fine solid particles that are separated and entrained in the fireball may be less than 0.1 Tg/Mt, regardless of whether the material is strong rock or unconsolidated particles.

As height of burst is increased, delivery of energy to the shock in the ground drops rapidly. The principal sources of dust become particles condensed from vapor and particles swept up from the surface. At sufficiently low height of burst, some surface material would be completely vaporized by radiation from the early fireball and later would condense to fine particles as the fireball cooled. At greater distances, only water and other relatively volatile constituents would be vaporized by optical photons from the fireball. The gas thus produced would loft solid particles and melt droplets into the fireball.

Finally, as the fireball rose, the afterwinds would scour the surface. This scouring could be an important source of dust if a dry, fine particulate soil were present at the target or if previous bursts had dried, crushed, and loosened the soil and raised precursor dust clouds.

The variation in mass loading of the fireball as the height of burst is increased is illustrated schematically in Figure 4.4.

In conclusion, materials directly vaporized by the nuclear explosion as well as ejecta melt are the principal sources of the fine particles lofted by nuclear clouds. Because these processes are relatively insensitive to soil and rock type, data from high-yield explosions on coral islands can reasonably be used to estimate the dust lofted by continental bursts.

These considerations of source mechanisms suggest that the mass of particulates lofted to stabilization altitude by surface bursts would be a few times 0.1 Tg/Mt. The available data will be examined next.

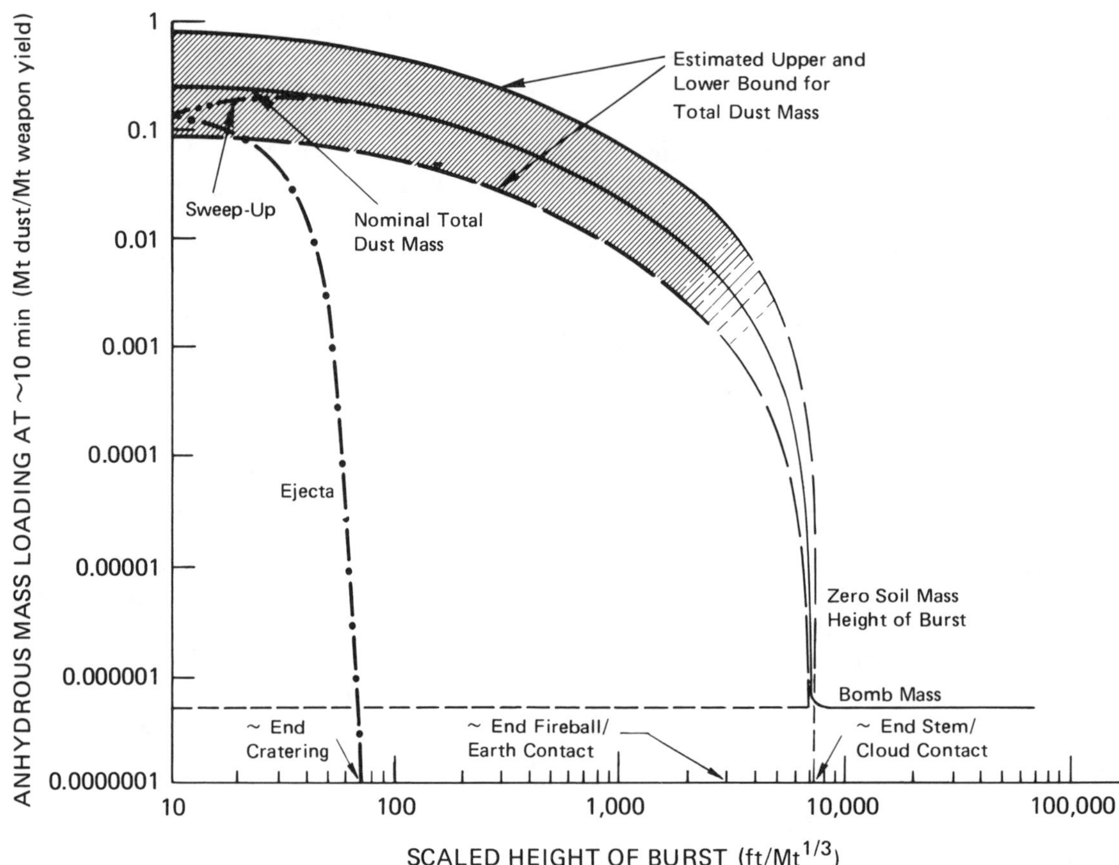

FIGURE 4.4 Dust mass per megaton lofted into the stabilized cloud as a function of height of burst showing relative contributions of blast ejecta and sweep up for bursts on or over moderately dry desert alluvium. (Provided by J. Carpenter, R&D Associates; now with Carpenter Research Corporation.)

OBSERVATIONS OF NUCLEAR DUST CLOUDS

The number of primary sources for data about the particles lofted to stabilization altitude by nuclear clouds is small. Early work is reviewed by Bjornerstedt and Edvarson (1963). More recent analyses include an important series of papers by Nathans et al. (1970), Heft (1970), and Gutmacher et al. (1983). Much of the early work on mass lofting was done with fallout samples, because radioactive fallout was a concern that demanded immediate attention. The early studies (e.g., Adams and O'Conner, 1957) elucidated the types and origins of particles in the fallout samples. Unfortunately, fallout samples are generally not representative of the particle distribution in the stabilized cloud.

Even in the early days of atmospheric nuclear weapons testing, samples of bomb debris were collected from the nuclear clouds with both manned and unmanned drone aircraft. The latter method was superior because it permitted sample collection as early as 30 min to 6 h after the explosions. Other programs aimed at collecting samples from domestic and non-U.S. tests probed clouds from 24 h to several days after the tests. In all these programs the emphasis was on obtaining representative samples throughout the cloud as well as minimizing the effects of fractionation (variation of specific radioactivity with particle size). Typically, the early (hours) samples contained one part in 10^{10} or 10^{11} of the original radioactivity. These samples were, nonetheless, "hot." The late samples (days) contained 10 to 100 times less. Generally, two types of debris-sampling systems were used. On the early (hours) missions, a single sample was collected per sortie; multiple samples per sortie were collected on late (days) missions. Both systems collected complete samples; the sampling mechanisms were designed with aerodynamic characteristics that minimized the possibility of deflection of small particles away from the samplers. The different techniques were made necessary by the high radioactivity at early times; both sampling systems used the same filter medium (IPC 1478), consisting of a cellulose bed on a cotton scrim backing.

The samples were subjected to several modes of analysis, including direct observation of the particles with electron microscopes, activation analysis, spectroscopy, observations of thin sections (made from the larger particles), and sedimentation analysis. Critical for the present report is the latter type of analysis, which yields information on the particle size distribution. Unfortunately, the sedimentation analysis, using a liquid medium, has two serious drawbacks: (1) the liquid tends to disassemble aggregated particles, thus increasing the fine component of the sample, and (2) the collection and counting scheme becomes increasingly inefficient at smaller particle sizes, thus decreasing the apparent fine particle component.

PARTICLE SIZE DISTRIBUTIONS

The principal results concerning particle size distribution are to be found in Nathans et al. (1970):

1. The total lofted mass is about 0.2 to 0.5 Tg/Mt.
2. For particles smaller than a few microns, the size distribution is roughly log normal with a mean radius (r_m) between 0.15 and 0.35 μm and a standard deviation (σ) of about 2 ± 0.5.
3. Above a few microns the size distribution is a power function with an exponent (α) probably bounded by 3.5 and 4.2.

It should be pointed out that the Nathans et al. data involve samples from only three surface nuclear tests: Johnie Boy, 0.5 kt, Nevada; Koon, 100 kt, Bikini; and Zuni, 3500 kt, Bikini. The observed dust number and mass distribution are quite variable, and size distribution parameters have been deduced from these data only in a qualitative sense.

Turco et al. (1983) define the size distribution,

$$M(r) = \frac{4\pi}{3}\rho r_m n_0 \left(\frac{r_0}{r_m}\right)^{\frac{\alpha + 1}{2}} \int_0^{\min(r,r_0)} x^2 \exp\left[-\frac{1}{2}\frac{\ln(x/r_m)^2}{\ln \sigma}\right] dx$$

$$+ \left(\frac{4\pi}{3}\rho n_0 r_0^{\alpha}\right) \int_{\min(r,r_0)}^{r} x^{3-\alpha}\, dx, \qquad (4.3)$$

where M(r) is the mass in particles smaller than radius r; r_0^α is the transition radius separating the log normal and power function portions of the size distribution; n_0 is a normalization constant; and ρ is the grain density. When normalized, equation (4.3) can be integrated to give the mass fraction of particles smaller than 1-μm radius ("submicron mass fraction"). Because these are the only particles with significant optical cross sections and long atmospheric lifetimes, the submicron fraction is an important characteristic of the size distribution. Figure 4.5 illustrates the submicron fraction (from equation (4.3)) as a function of the mean radius (r_m) of the log normal part of the distribution, for a set of plausible values of the standard deviation (σ) and the exponent (α) of the power law segment. These curves illustrate the large sensitivity of the "optical" impact of the dust to the size distribution parameters. For fixed values of the width (σ) and exponent (α), the submicron mass fraction increases with decreasing mean particle radius (r_m). For a fixed mean radius, either increasing the exponent or decreasing the width increases the submicron mass fraction.

The nuclear dust data summarized above suggest that the family of curves illustrated in Figure 4.5 defines a plausible range for the submicron fraction. The integrations of equation (4.3) were cut off at 1-cm radius, because the nuclear clouds could not lift larger particles. A nominal value for the mass fraction of particles smaller than 1-μm radius is 8 percent, with a range from a few percent to perhaps 20 percent.

OPTICAL PROPERTIES OF AIRBORNE DUST

The optical properties of dust raised in nuclear clouds have not been directly determined. However, much information is available for volcanic dust and for natural windblown dust. The former type of dust may represent the glassy, partly melted fraction in a nuclear cloud, and windblown dust should be similar to dust swept up by the rising fireball.

The visible wavelength optical properties of volcanic dust have most recently been reviewed by Patterson et al. (1983). They estimate that the 500-nm refractive index for the stratospheric dust from El Chichon's 1982 eruption was n = 1.53 - 0.001 i, based on measurements of dust collected at the ground 80 km from the volcano. The imaginary part of the refractive index was nearly independent of wavelength from

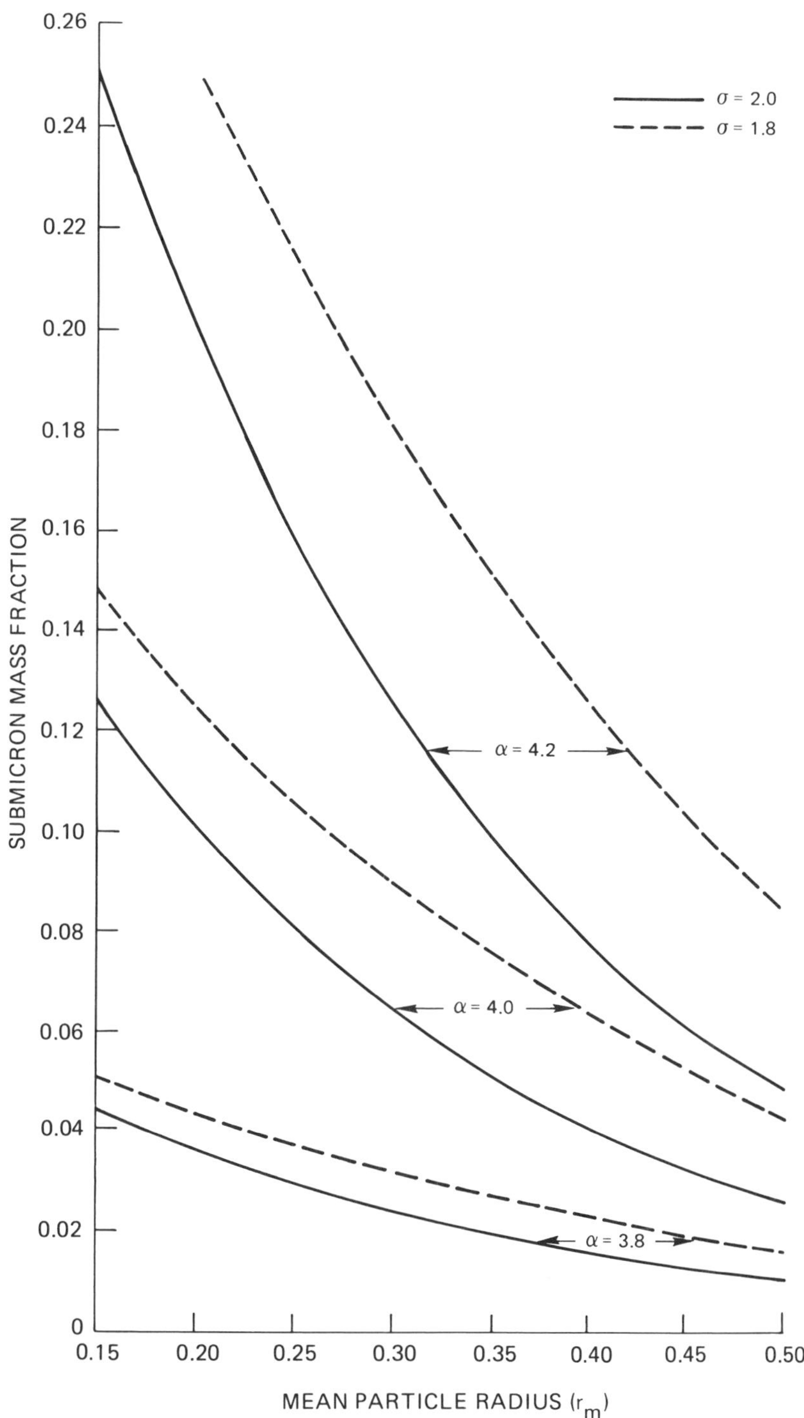

FIGURE 4.5 Estimates of the fraction of dust smaller than 1-μm radius for values of the mean radius (r_m), the width (σ) of the log normal portion of the size distribution, and the exponent (α) describing the power law portion of the distribution. Nominal values of the parameters ($\alpha = 4$, $\sigma = 2$, and $r_m = 0.25$) give a submicron mass fraction of 8 percent.

350 to 700 nm and increased slightly below 350 nm. By contrast the imaginary index for Mount St. Helens dust was three times larger. The range of real indices of refraction for various volcanic glasses is about 1.48 to 1.63 (Patterson et al., 1983) and may vary systematically with SiO_2 content of the rock.

Pollack et al. (1973) measured the visible and infrared optical constants of a variety of volcanic and crustal rocks. They found a real index of refraction in the visible between 1.47 and 1.57. The imaginary indices of refraction ranged from 2×10^{-5} for pure volcanic glasses such as obsidian to 1×10^{-3} for a crustal rock such as andesite.

Patterson (1981) has summarized the visible optical constants of a number of crustal rocks. The imaginary index tends to decline as particle size increases above 10 μm, because the quartz fraction increases. Smaller clay particles are reddish in color, and the absorption is probably due to iron, which can vary widely in form and concentration. Generally, the imaginary index at 500 nm is between 5×10^{-3} and 1×10^{-2} and varies with wavelength (λ) as strongly as λ^{-3} from 300 to 700 nm.

At infrared wavelengths, all volcanic and crustal rocks show considerable wavelength dependence due to the strong silicate bands located near 10 μm. Unfortunately, Patterson (1981) does not report data beyond 20 μm, but Pollack et al. (1973) and Toon et al. (1977) have shown that there are also strong bands just beyond 20 μm. At some infrared wavelengths the imaginary optical constants may differ by an order of magnitude between various rock types. However, near the 8- to 12-μm band centers, which are the crucial wavelengths since they lie in the atmospheric window, the measurements appear to be within a factor of 2 or 3 for different rock types and different measurement techniques.

Turco et al. (1983) employed a visible refractive index of 1.5 - 0.001 i for dust, which agrees well with the values found for volcanic dust. However, this imaginary index is somewhat lower than that of typical crustal debris. Since the committee concludes that the majority of the fine dust in the stabilized cloud would be vaporized or melted, it also adopts as most appropriate the refractive index of 1.5 - 0.001 i for all visible wavelengths. Turco et al. (1983) used the optical constants of basaltic glass (Pollack et al., 1973) in the infrared. These optical constants are at the high end of the range of the imaginary index at 10 μm for a variety of crustal materials but seem best suited for glassy ejecta, so the committee also adopts them in this study.

DUST LOFTED IN THE BASELINE CASE

The 6500-Mt baseline case includes 400 weapons of 1 Mt or greater and 2000 smaller weapons averaging 0.5 Mt detonated as surface bursts, presumably against hard targets such as silos and buried command structures.

TABLE 4.1 Dust Lofting[a] at Mid-Latitudes in the Baseline Case

	1.5 Mt	1 Mt	0.5 Mt	Total
Weapons	200	200	2000	2400
Yield (Mt)	300	200	1000	1500
Cloud dust (Tg)	60-150	40-100	200-500	300-780
Stem dust[b] (Tg)	6-15	4-10	20-50	30-75
Total dust (Tg)	66-165	44-110	220-550	330-825
Cloud top (km)	19	17	14	--
Cloud bottom (km)	9	8	7	--
Tropopause (km)	12	12	12	--
Stratosphere dust (Tg)	42-105	22-55	57-143	121-303
Tropospheric dust[c] (Tg)	24-60	22-55	163-407	209-522
Stratospheric submicron[d] dust (Tg)	3-8	2-4	5-11	10-23
Tropospheric submicron[d] dust (Tg)	2-5	2-4	13-33	17-42

[a]Dust values are shown for lofting efficiencies of 0.2 Tg/Mt (first value) and 0.5 Tg/Mt (second value).

[b]Stem dust is assumed to be 10 percent of cloud dust, based on comparison of stem and cloud volumes. This is probably a significant underestimate of the dust loading in the lower parts of the stem but more accurate in the top of the stem, which contributes to the stratospheric dust totals. No adequate data base exists to reduce this uncertainty; however, stem dust is not expected to be a major factor in the global climatic effect.

[c]The tropospheric dust is the sum of the stem dust and the portion of the cloud dust below the troposphere.

[d]The submicron mass is calculated assuming 8 percent of the dust mass lies in the submicron size range.

At mid-latitudes the dust lofting is characterized by parameters shown in Table 4.1. The figures are based on the assumptions that the cloud dust is uniformly distributed from the cloud top to the bottom, that the stem contains 10 percent of the cloud dust, and that the submicron mass fraction is 8 percent. The committee recognizes that these assumptions ignore the likely variation of air and dust density with altitude. More detailed estimates could be obtained from computer simulations. However, given the uncertainties in mass lofting and source description, such detail seems unwarranted. Scenarios leading to much larger dust effects would require more detailed treatment. The ranges of lofted dust are assumed to arise only from the plausible range of 0.2 to 0.5 Tg/Mt for the lofting capabilities of the nuclear clouds. The most probable value of the lofted dust is 0.3 Tg/Mt, resulting in an estimated 15 Tg of stratospheric submicron dust. If the uncertainty in the submicron dust fraction is included, the overall range of uncertainty of potential dust injections increases further.

EXCURSIONS

The mass of submicron dust lofted into the stratosphere in the baseline case is relatively small (10 to 24 Tg) in comparison with masses in the cases studied by Turco et al. (1983). Contributing to this difference are the smaller weapon yields and the reduced total megatonnage in surface bursts that have been assumed in the baseline case.

The committee considered excursions that might increase the role of dust in postwar climatic effects. The main, 8500 Mt, excursion adds 100 20-Mt surface bursts that might be used in attacks on superhard targets. The clouds from such bursts would reach 37 km (top) and 19 km (bottom), so that virtually all the lofted dust would reach the stratosphere. The lofted mass would be 400 to 1000 Tg (600 Tg likely), with 8 percent of the mass in the submicron fraction.

The committee also considered a simultaneous attack totaling 500 Mt of surface bursts against a cluster of closely spaced hard targets. As discussed earlier, the rise of the resulting giant fireball would be qualitatively different from the rise of single-megaton buoyant fireballs. The rise rates are much greater (kilometers per second, instead of 100 m/s), so that the lofting efficiency might exceed the energy-constrained maximum of 2.6 Tg/Mt expected for buoyant fireballs. For example, the impact proposed by Alvarez et al. (1980, 1982) to explain the iridium-enriched Cretaceous-Tertiary (K-T) boundary claystone apparently lofted a total of 10^7 Tg (10^{19} g) of dust. If the 10-km diameter impactor had a velocity of 30 km/s, its kinetic energy would have been about 10^8 Mt. Most of this energy was deposited in the target material, but perhaps 5 percent (5×10^6 Mt) appeared as thermal energy of the vaporized projectile and target material (Jones and Kodis, 1982). The explosive expansion of this high-pressure gas created an enormous fireball that was unconfined by the atmosphere and probably provided the energy to spread the dust worldwide. The implied lofting efficiency of the Cretaceous-Tertiary fireball is roughly 2 Tg/Mt. If this efficiency is used for the 500-Mt fireball in the postulated simultaneous attack, the mass lofted to very great altitude (perhaps 100 km; C.E. Needham, S-Cubed, Inc., Albuquerque, unpublished numerical simulations of 500-Mt explosions, 1982) would be about 1000 Tg. This value is comparable with the dust lofted by the 100 20-Mt bursts in the 8500-Mt excursion.

SUMMARY

The mass of submicron dust lofted into the stratosphere during a nuclear war would depend most critically on the following factors: (1) the number and individual yields of weapons used in surface bursts, (2) the lofting efficiency of the fireballs, and (3) the size distribution of particles in the stabilized cloud. Available data confirm that the size distribution is roughly log normal below a few microns (r_m = 0.25 μm, σ = 2) and follow a power law (α = 4) at larger sizes. The lofting efficiency is probably 0.3 Tg/Mt (within an observed range of 0.2 to 0.5) for yields capable of reaching the stratosphere. These estimates agree well with those of Turco et al. (1983) for size distribution and lofting efficiency.

For the present 6500-Mt baseline case the total mass of lofted dust would be of the order of 500 Tg, with a stratospheric submicron total of 15 Tg. Within the baseline case the latter figure is unlikely to exceed 30 Tg.

An excursion involving higher yield weapons or concentrated attacks on hard targets might increase the masses to as much as 1000 Tg (total) and 80 Tg (stratospheric submicron particles).

REFERENCES

Adams, C.E., and J.D. O'Connor (1957) The Nature of Individual Radioactive Particles. VI. Fallout Particles from a Tower Shot, Operation Redwing. Report NRDL-TR-208. San Francisco, Calif.: U.S. Naval Radiological Defense Laboratory.

Alvarez, L.W., W. Alvarez, F. Asaro, and H.W. Michael (1980) Extraterrestrial cause for the Cretaceous-Tertiary extinction. Science 208:1095-1108.

Alvarez, W., L.W. Alvarez, F. Asaro, and H.W. Michael (1982) Current status of the impact theory for the terminal Cretaceous extinction. Geol. Soc. Am. Spec. Pap. 190:305.

Bethe, H.A., K. Fuchs, J.O. Hirschfelder, J.L. McGee, R.E. Peierls, and J. von Neumann (1947) Blast Wave. Report LA-2000. Los Alamos, N.Mex.: Los Alamos Scientific Laboratory.

Bjornerstedt, R., and K. Edvarson (1963) Physics, chemistry, and meteorology of fallout. Annu. Rev. Nucl. Sci. 13:505.

Brode, H.L. (1955) Numerical solutions of spherical blast waves. J. Appl. Phys. 26:766.

Brode, H.L. (1968) Review of nuclear weapons effects. Annu. Rev. Nucl. Sci. 18:153.

Butkovich, T.R. (1974) Rock Melt from an Underground Nuclear Explosion. Report UCRL-51554. Livermore: University of California, Lawrence Livermore National Laboratory. 10 pp.

Cooper, H.F. (1977) A summary of explosion cratering phenomena relevant to meteor impact events. Pages 11-44 in Impact and Explosion Cratering, edited by D.J. Roddy, R.O. Pepin, and R.B. Merrill. New York: Pergamon Press.

EG & G Technical Staff (1958) Operation Castle: Cloud Photography. Report WT-933. Boston, Mass.: U.S. Atomic Energy Commission.

Foley, H.M., and M.A. Ruderman (1973) Stratospheric NO production from post-nuclear explosions. J. Geophys. Res. 78:4441.

Gault, D.E., E.M. Shoemaker, and H.J. Moore (1962) Spray ejected from the lunar surface by meteoroid impact. NASA Tech. Note D-1767. 39 pp.

Glasstone, S., and P.J. Dolan (1977) The Effects of Nuclear Weapons (3rd edition). Washington, D.C.: U.S. Government Printing Office.

Gutmacher, R.G., G.H. Higgins, and H.A. Tewes (1983) Total Mass and Concentration of Particles in Dust Clouds. Report UCRL-14397 Revision 2. Livermore: University of California, Lawrence Livermore National Laboratory.

Heft, R.E. (1970) The characterization of radioactive particles from nuclear weapons tests. Radionuclides in the Atmosphere. Adv. Chem. Ser. 93:254.

Horak, H.G., and J.W. Kodis (1983) RADFLO--A User's Manual. Report LA-9245. Los Alamos, N.Mex.: Los Alamos National Laboratory.

Horak, H.G., E.M. Jones, M.T. Sandford II, R.W. Whitaker, R.C. Anderson, and J.W. Kodis (1982) Two-Dimensional Radiation-Hydrodynamic Calculation of a Nominal 1-Mt Nuclear Explosion Near the Ground. Report LA-9137. Los Alamos, N.Mex.: Los Alamos National Laboratory.

Jones, E.M., and J.W. Kodis (1982) Atmospheric effects of large-body impacts: The first few minutes. Geol. Soc. Am. Spec. Pap. 190:175.

Jones, E.M., and M.T. Sandford (1977) Numerical simulations of a very large explosion at the earth's surface with possible applications to tektites. Page 1009 in Impact and Explosion Cratering. New York: Pergamon Press.

Melosh, H.J. (1982) The mechanics of large meteoroid impacts in the earth's oceans. Geol. Soc. Am. Spec. Pap. 190:121.

Moore, H.J., and D.E. Gault (1965) The fragmentation of spheres by projectile impact. Pages 127-164 in Astrogeologic Studies Annual Report, July 1, 1964 to July 1, 1965, Part B: Crater Investigations. Flagstaff, Ariz.: Department of the Interior, U.S. Geological Survey.

Nathans, M.W. (1970) The specific activity of nuclear debris from ground surface bursts as a function of particle size. Radionuclides in the Atmosphere. Adv. Chem. Ser. 93:352.

Nathans, M.W., and R. Thews (1970) The particle size distribution of nuclear cloud samples. Radionuclides in the Atmosphere. Adv. Chem. Ser. 352:360.

Nathans, M.W., R. Thews, W.D. Holland, and P.A. Benson (1970) Particle size distribution in clouds from nuclear airburst. J. Geophys. Res. 75:7559.

National Research Council (1975) Long-Term Worldwide Effects of Multiple Nuclear Weapons Detonations. Washington, D.C.: National Academy of Sciences.

Öpik, E.J. (1958) Physics of Meteor Flight in the Atmosphere. Interscience Tracts of Physics and Astronomy 6. New York: Interscience.

Patterson, E.M. (1981) Optical properties of the crustal aerosol in relation to chemical and physical characteristics. J. Geophys. Res. 86:3236-3246.

Patterson, E.M., C.O. Pollard, and I. Galindo (1983) Optical properties of the ash from El Chichon volcano. Geophys. Res. Lett. 10:317-320.

Peterson, K.R. (1970) An empirical model for estimating world-wide deposition from atmospheric nuclear detonations. Health Phys. 18:357-378.

Pollack, J.B., O.B. Toon, and B.N. Khare (1973) Optical properties of some terrestrial rocks and glasses. Icarus 19:372-389.

Sedov, L.I. (1959) Similarity and Dimensional Methods in Mechanics. New York: Academic Press.

Sowle, D.H. (1975) Implications of Vortex Theory for Fireball Motion. Report DNA-3581F. Washington, D.C.: Defense Nuclear Agency.

Taylor, G.I. (1950) The formation of a blast wave by a very intense explosion. I. Theoretical discussion. Proc. Roy. Soc. Ser. A 2:159.

Toon, O.B., J.B. Pollack, and C. Sagan (1977) Physical properties of the particles composing the Martian dust storm of 1971-1972. Icarus 30:663-696.

Turco, R.P., O.B. Toon, T.P. Ackerman, J.B. Pollack, and C. Sagan (1983) Global Atmospheric Consequences of Nuclear War. Interim Report. Marina del Rey, Calif.: R&D Associates. 144 pp.

Zinn, J. (1973) A finite difference scheme for time-dependent, spherical radiation hydrodynamics problems. J. Comp. Phys. 13:569.

5
Fires

OVERVIEW*

It is clear that nuclear explosions can ignite large-scale fires (Broido, 1960).† In addition, it has been estimated that the smoke emissions from nuclear-initiated fires could produce major atmospheric perturbations (Lewis, 1979; Crutzen and Birks, 1982; Turco et al., 1983a,b). Only two nuclear explosions have ever occurred over populated areas (Hiroshima, August 6, 1945, and Nagasaki, August 9, 1945); in each case, a city-sized conflagration resulted. At Hiroshima, a ≃12-kt weapon caused a mass fire over an area of ≃13 km^2, essentially the entire central city (Ishikawa and Swain, 1981). At Nagasaki, where high terrain shadowed large regions of the city from direct irradiation by bomb light, a ≃20-kt device burned ≃7 km^2 (Ishikawa and Swain, 1981). It is difficult to extrapolate the effects of these two isolated events, which involved <40-kt total yield, to the possible effects of a global nuclear exchange involving 6500 Mt. Nevertheless, a logical sequence of steps can be taken to obtain estimates of the areal extent and particulate emissions of fires initiated in a full-scale nuclear war:

1. Review historical fire experience to assess the probability of ignition and spread of large fires.
2. Define the effectiveness of nuclear explosions for initiating fires in urban and forest settings.

---

*In the text, the following symbols are used: ≃, approximately equal to; ~, of the order of; $\lesssim$, less than or of the order of; $\gtrsim$, greater than or of the order of.

†For the purposes of this report, large-scale fires can be classified as "mass fires," in which many individual fires burn simultaneously over a large area, "conflagrations," in which the fire is most intense along a line of propagation, "firestorms," in which the entire area of the fire burns intensely and strong winds blow inward from all directions, and "fire whirls," in which a firestorm plume develops an unusually strong vorticity.

3. Determine the burdens and distributions of combustible materials around potential nuclear targets.

4. Evaluate data on the quantity and physical properties of smoke generated by common fuels.

5. Consider mass fire dynamics to determine the likely heights and rates of injection of the smoke.

6. Describe a scenario for the locations, yields, and heights of nuclear detonations (See Chapter 3, "The Baseline Nuclear Exchange").

7. Combine the foregoing information to estimate the total quantity and optical characteristics of nuclear war smoke emissions.

These topics are discussed in subsequent sections of this chapter. On the basis of such an analysis, an approximate equation can be written that emphasizes the important factors that enter into the estimation process,

$$E = Y_f A_0 m_0 f_b \varepsilon \times 10^{10},$$

where E is the total smoke emission (in grams), Y_f is the total explosion yield (in megatons) in air bursts that effectively ignite fires, A_0 is the average area ignited by each megaton of yield (in square kilometers per megaton), m_0 is the average loading of flammable materials (in grams per square centimeter), f_b is the fraction of m_0 burned, and ε is the mean smoke emission factor (grams of smoke per gram of material burned). The factor of 10^{10} converts square kilometers (A_0) to square centimeters.

The key parameter values that apply to the baseline nuclear war scenario are given in Table 5.1. The total smoke emission calculated for the baseline case is ≈180 Tg (1 Tg = 10^{12} g ≈ 10^6 metric tons), or ≈0.7 g/m^2 averaged over the northern hemisphere. Since the specific extinction (scattering plus absorption) coefficient of many smokes at visible wavelengths is ≈5.5 m^2/g, the hemispherical average optical depth* in this case is ≈4. Of course, if the smoke were confined to the northern mid-latitude zone, the optical depth would be ≈2 to 3 times larger, or ≈8 to 12. A more detailed discussion of these estimates follows. The optical and climatic effects of the smoke are discussed in Chapter 7.

PRESENT-DAY SMOKE EMISSION AND REMOVAL

It is estimated that the current global smoke emission to the atmosphere is ≈200 Tg/yr (Seiler and Crutzen, 1980; Turco et al., 1983a,c). The graphitic carbon fraction is about 5 to 10 percent by

---

*The optical depth is a dimensionless quantity that determines the light transmission properties of a layer of gas or aerosols. If the layer has an optical depth τ, $e^{-\tau}$ is the fraction of a beam of light perpendicularly incident on the layer that suffers no scattering or absorption in passing through the layer. The total light transmitted consists of the direct light plus a scattered (diffuse) component.

TABLE 5.1 Baseline Nuclear War Fire and Smoke Parameters[a]

Parameter[b]	Urban Fires	Forest Fires
Y_f (Mt)	1000	1000
A_0 (km^2/Mt)	250	250
m_0 (g/cm^2)	4	2
f_b	0.75	0.20
ε (g/g)[c]	0.02	0.03

[a]Excursions from the baseline case, and uncertainties in the baseline parameters, are discussed in the text.
[b]Y_f is the effective ignition yield in megatons, A_0 is the average ignition area per megaton, m_0 is the burden of combustibles per unit area, f_b is the fraction of the combustibles burned, and ε is the net smoke emission factor per unit of fuel, assuming in the case of urban fires that 50 percent of the smoke is promptly scavenged and removed from the plumes mainly as "black rain."
[c]The smoke consists of 20 percent graphitic carbon (soot) by mass, and 80 percent transparent oily compounds.

NOTE: Urban fire smoke emission: $E_u = 150 \times 10^{12}$ g
Forest fire smoke emission: $E_f = 30 \times 10^{12}$ g
Total smoke emission: $E_t = 180 \times 10^{12}$ g

mass. The primary sources of smoke are agricultural burning, fossil fuel combustion, and wildfires. The important characteristics of background smoke emissions that distinguish them from "nuclear" fire emissions are as follows:

1. The smoke emission factors are low in relation to the quantity of fuel burned, because most of the burning takes place under controlled conditions.
2. The overall graphitic carbon component is low, because most of the smoke is generated during the prescribed combustion of natural cellulosic materials.
3. Almost all of the smoke is injected into the lowest 1 km of the atmosphere, because the sources are small in horizontal scale and/or total power.
4. The smoke emissions occur in diverse locations throughout the course of a year, which prevents significant concentrations from building up.
5. The average atmospheric lifetime of the smoke is $\lesssim 10$ days (Ogren, 1982).

As a result of these factors, the average background concentration of airborne graphitic carbon is typically only <0.1 μg/m^3, and its integrated vertical absorption optical depth is <0.01 (Charlson and Ogren, 1982; Turco et al., 1983c). Over a period of about 1 month, background smoke emissions would be negligible in comparison with the estimated smoke emissions of a nuclear war (Turco et al., 1983a,b; Crutzen et al., 1984).

Removal of smoke and soot from the atmosphere occurs mainly through precipitation scavenging. Smoke particles have sizes of about 0.1- to 0.5-μm radius, at which sedimentation is negligible and dry deposition is very inefficient (Slinn, 1977; Sehmel, 1980). In the background atmosphere, soot is usually found as a minor component of hygroscopic sulfate aerosols. This suggests removal by efficient scavenging of the hygroscopic aerosols in and below clouds (Radke et al., 1980; Ogren, 1982; Turco et al., 1983c).

The arctic haze that forms in winter and spring is known to contain soot (Rosen and Novakov, 1983). The haze is (relatively) highly absorbing because of the soot it holds (Patterson et al., 1982). The seasonal conditions that lead to the formation of the winter polar vortex create a stable air mass with low precipitation in which carbon emissions produced by combustion can remain suspended for several months. This demonstrates that under some meteorological conditions, particularly with the suppression of precipitation, smoke and soot can have an extended atmospheric lifetime.

Generally speaking, it is expected that smoke from nuclear-initiated fires would have a longer atmospheric lifetime than background smoke (notwithstanding prompt scavenging in the fire plumes), because of its greater heights of injection. This point is expanded in subsequent sections.

HISTORICAL FIRE EXPERIENCE

Human experience with mass fires and firestorms includes urban conflagrations triggered by natural disasters (e.g., earthquakes), wartime city fires initiated by incendiary and nuclear bombing, massive wildfires and forest fires, and field experiments with large-scale fuel beds (Carrier et al., 1982). Although few of these experiences are directly applicable to the nuclear war problem, all contribute to a general understanding of the properties and behavior of large-scale fires.

Earthquakes

Earthquakes have started urban conflagrations by breaking gas lines, exposing stored fuels, shorting electrical circuits, breaching open fires, and hampering effective firefighting. Particularly striking examples of large fires induced by earthquakes occurred in San Francisco in 1906 and Tokyo in 1923. A nuclear blast wave would have a similar impact and, in combination with the thermal light pulse, would represent a much greater fire threat than an earthquake.

World War II

The World War II saturation bombing of German and Japanese cities provided ample evidence that mass fires can be readily ignited in urban settings. The nuclear explosions over Hiroshima and Nagasaki are discussed later. The conventional bombing of cities such as Hamburg, Dresden, Darmstadt, and Tokyo produced intense fires over many square kilometers and, in some instances, triggered firestorms. From anecdotal evidence, it is known that thick, dark plumes rose from these fires to altitudes of 6 to 12 km. Within the fire zones, almost all the buildings were gutted and all combustible materials consumed. Such experiences show that, when many simultaneous fire ignitions occur among closely spaced structures and firefighting capability is suppressed, mass fires are likely to develop.

Occasionally, massive urban conflagrations, such as the Great Chicago Fire of 1871, are touched off by single ignitions (Kerr, 1971). Although such fires are not typical, they are symptomatic of the hazardous fire conditions that exist in many crowded urban centers.

Forest Fires

Plummer (1912), Ayers (1965), and F.E. Fendell (in Appendix 5-1), among others, have reviewed the largest forest fires of the past 160 years in which areas up to 20,000 km^2 were blackened. The conditions under which these catastrophic fires developed included long drought, low humidity, and high winds (e.g., Plummer, 1912). Clearly, such conditions are not common over large areas of the northern hemisphere during most of the year (Chandler et al., 1963). However, for the analysis of nuclear-induced fires, three general types of fire danger conditions should be distinguished: (I) fires are difficult to ignite and do not spread if ignited; (II) fires are readily ignited, but their spread is limited by factors such as humidity, moisture, topography, winds, and firebreaks; and (III) fires readily ignite and spread uncontrollably over large areas.

Historical catastrophic forest fires are exclusively of type III. By contrast, most nuclear forest fires would probably be of type II. Historical fires are characterized by a limited number of ignition points, perhaps one ignition for each 50 to 500 km^2 burned (Ayers, 1965). Nuclear explosions, by contrast, can ignite forest debris instantly over a large area, with numerous ignition points developing into moderate size fires (although the probability of extensive fire spread outside of the original burning zone would be much lower--see below).

The great Tunguska meteor, which fell over Siberia on June 30, 1908, provides a very rough indication of the effects that might be produced by a high-yield nuclear explosion over a forest. The Tunguska event was equivalent, in terms of the blast wave, to a ≈10-Mt detonation at ≈8-km altitude (Krinov, 1966). (As noted below, high-yield nuclear bursts have smaller incendiary efficiencies than low-yield bursts.) Roughly 1600 km^2 of Siberian forest was flattened. Eyewitness accounts describe "burning falling trees" and widespread fires. A series of Russian scientific expeditions to the fall site concluded that several major fires had broken out in the central zone of devastation and burned for 5 days. From the description of the charred remains, it appears that bark and many small branches were stripped from the trees and burned, to an extent not usually observed in natural fires of that area.

Experimental Fires

Experimental large-scale fires have been used to study fire development and plume dynamics. Among these experiments are the Flambeau series (Martin, 1974; Palmer, 1981), the Euroka fires (Williams et al., 1970), and the Meteotron events (Dessens, 1962; Church et al., 1980). However, because the extent of these fires was only about 10^3 to 10^5 m^2, extrapolation of the results to city-size fires is difficult. Of particular interest here is the height of the smoke plume in a large fire. In the experiments noted above, the plume aspect ratio (i.e., the plume height divided by the fire diameter) was always >>1, and the plumes often formed vortices penetrating to heights >1 km. (The plume aspect ratio cannot be simply scaled to larger fires. The dependence of plume height on fire size and intensity, and extrapolations to city-sized fires are discussed in a later section.)

The Flambeau experiments also led to the definition of a set of conditions for firestorm genesis that has been widely accepted (FEMA, 1982). The conditions include a fuel loading of >4 g/cm^2, a building density of ≥20 to 30 percent, a fire area of >3 km^2, initial fires in >20 percent of the buildings, and ambient winds of <10 km/h (Baldwin, 1968; Martin, 1974). However, these conditions are still controversial, as they have never been tested on an appropriate scale. Moreover, in view of the atmospheric effects being considered here, it is not clear that firestorms and very intense mass fires need to be differentiated, except perhaps to refine the estimation of smoke injection altitudes (see below).

IGNITION OF NUCLEAR FIRES

Thermal Phenomena

In a nuclear air burst at low altitude (<10 km), about 30 to 40 percent of the energy is released as an intense pulse of visible light; about 45 to 55 percent of the energy is converted to blast pressure waves; and about 15 percent is contained in prompt and delayed nuclear radiation (Glasstone and Dolan, 1977, hereafter GD77). Most of the

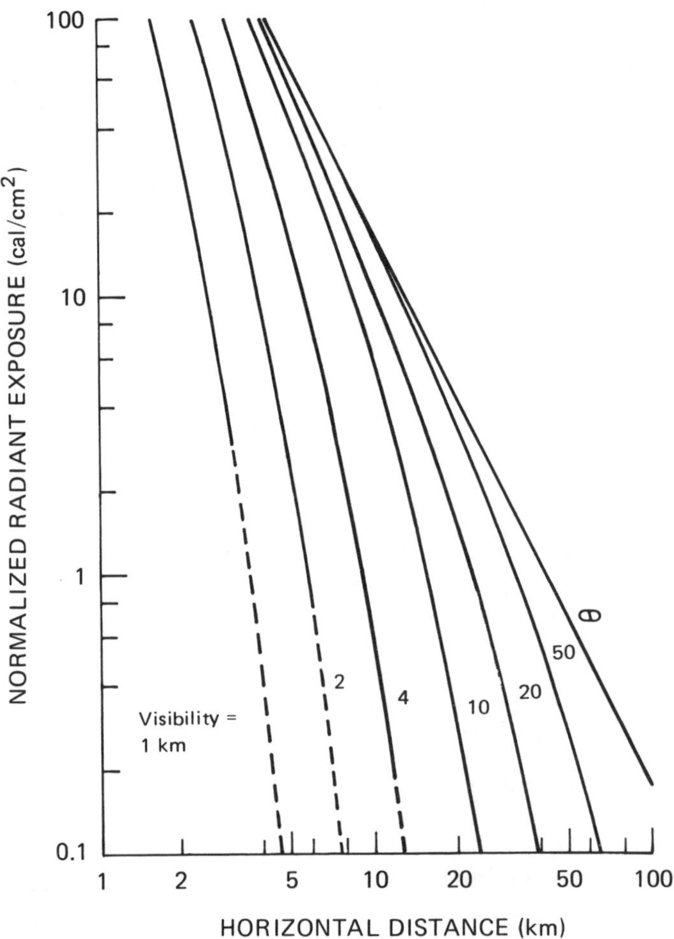

FIGURE 5.1 Maximum radiant exposures versus ground range from a 1-Mt air burst (detonated below several kilometers altitude) as a function of the ground level visibility. The radiant exposures scale roughly with the yield in megatons. (From Kerr et al., 1971)

bomb light is emitted within a few seconds for megaton yield explosions, and in less than a second for kiloton-size bursts (GD77). For a 1-Mt low air burst, Figure 5.1 shows the thermal fluences (in calories per square centimeter incident on a surface normal to the line-of-sight through the burst point) as a function of distance from ground zero, and for various atmospheric visibilities. With a 1-Mt explosion and normal visibilities ($\gtrsim$10 km), the 20-cal/cm^2 thermal fluence contour lies about 7 km from the explosion hypocenter, versus 9 km in a perfectly transparent atmosphere. With a 100-kt explosion, atmospheric transmission, for visibilities of >5 km, has little effect on radiant exposures where fluences exceed 20 cal/cm^2. Lower visibilities restrict the range at which nuclear thermal effects are important. Oblique incidence of the bomb light on exposed surfaces also reduces the effective fluence. On the other hand, cloud and surface reflections enhance the radiant fluxes in localized regions.

As sunlight, focused by a lens, can ignite flammable materials, so can the thermal emissions of a nuclear explosion (Glasstone, 1957; Miller, 1962). Ignition data obtained during atmospheric nuclear test detonations and by laboratory experimentation are summarized in Table 5.2. At a specific thermal fluence, small nuclear explosions are generally more efficient at igniting fires than large explosions because the thermal pulse has a shorter duration and larger peak intensity (in addition, there is a lower probability of significant atmospheric attenuation over the shorter ranges involved). Newspaper, brown paper, cotton cloth, and dried plant material can be ignited by 10 cal/cm^2 from a $\lesssim$1-Mt explosion. The perimeter of the Hiroshima fire zone roughly coincided with the 10-cal/cm^2 contour. At Nagasaki, in directions unobscured by hills, the conflagration zone also extended roughly to the 10-cal/cm^2 limit.

In the application of nuclear weapons against "soft" targets (e.g., urban and industrial targets), peak overpressures* of $\gtrsim$5 psi (pounds per square inch) are often used to define the zone of assured destruction (GD77). The 5-psi contour circumscribes an area of $\approx$1.4 km^2/kt for a 1-kt explosion (at the optimum height of burst), $\approx$0.30 km^2/kt for a 100-kt explosion, and $\approx$0.14 km^2/kt for a 1-Mt explosion. The corresponding areas enclosed within the 20-cal/cm^2 thermal irradiance contours (GD77) are $\approx$0.30 km^2/kt, $\approx$0.30 km^2/kt, and $\approx$0.25 km^2/kt, respectively (in the 1-Mt case, the atmospheric visibility is assumed to be 20 km). In estimating the potential fire areas for nuclear air bursts, the committee has chosen an average ignition area of 0.25 km^2/kt (250 km^2/Mt) for individual explosions, which is roughly consistent with 5-psi overpressures and 20 cal/cm^2 thermal fluences at the limits of the ignition region, under normal conditions of atmospheric transmission. These areas are quite conservative in relation to the areas burned at Hiroshima ($\approx$1 km^2/kt) and Nagasaki ($\approx$0.35 km^2/kt).

The question of overlap of ignition zones for closely spaced detonations, and the total potential fire area in a full exchange, are discussed in a separate section of this chapter.

Close to the hypocenter of a nuclear explosion, the thermal energies are much larger than 20 cal/cm^2. Within the 30-cal/cm^2 contour (about 150 km^2 for a 1-Mt explosion), substantial quantities of natural and synthetic organic and cellulosic materials would be instantly pyrolized, and the combustible vapors ignited in a massive "flashover" fire. The rising fireball would then draw the flames and smoke toward the stem of the nuclear cloud, establishing the conditions for accelerated burning and, in some cases, the core of an incipient firestorm.

For surface and subsurface nuclear detonations, the potential thermal effects are greatly reduced (although the dust and prompt radioactive fallout effects are increased). The bomb light from a

---

*The term "overpressure" refers to the incremental static pressure above ambient atmospheric pressure (about 14.7 pounds per square inch at sea level) caused by the passage of the explosion wave.

TABLE 5.2 Approximate Radiant Exposures for Ignition of Various Flammable Materials for Low Air Bursts

Material	Color	Effect on Material	Radiation Exposure[a] (cal/cm²) 35 kt	1.4 Mt	20 Mt
Household Tinder Materials					
Newspaper, shredded		Ignites	4	6	11
Newspaper, dark picture area		Ignites	5	7	12
Newspaper, printed text area		Ignites	6	8	15
Crepe paper	Green	Ignites	6	9	16
Kraft paper	Tan	Ignites	10	13	20
Bristol board, 3 ply	Dark	Ignites	16	20	40
Kraft paper carton, used (flat side)	Brown	Ignites	16	20	40
New bond typing paper	White	Ignites	24[b]	30[b]	50[b]
Cotton rags	Black	Ignites	10	15	20
Rayon rags	Black	Ignites	9	14	21
Cotton string scrubbing mop (used)	Gray	Ignites	10[b]	15[b]	21[b]
Cotton string scrubbing mop (weathered)	Cream	Ignites	10[b]	19[b]	26[b]
Paper book matches, blue head exposed		Ignites	11[b]	14[b]	20[b]
Excelsior, ponderosa pine	Light yellow	Ignites	--[c]	23[b]	23[b]
Outdoor Tinder Materials[d]					
Dry rotted wood punk (fir)		Ignites	4[b]	6[b]	8[b]
Deciduous leaves (beech)		Ignites	4	6	8
Fine grass (cheat)		Ignites	5	8	10
Coarse grass (sedge)		Ignites	6	9	11
Pine needles, brown (ponderosa)		Ignites	10	16	21

TABLE 5.2 (continued)

Material	Color	Effect on Material	Radiation Explosure[a] (cal/cm^2) 35 kt	1.4 Mt	20 Mt
Construction Materials					
Roll roofing, mineral surface		Ignites	—[c]	>34	>116
Roll roofing, smooth surface		Ignites	—[c]	30	77
Plywood, Douglas fir		Flaming during exposure	9	16	20
Rubber, pale latex		Ignites	50	80	110
Rubber, black		Ignites	10	20	25
Other Materials					
Aluminum aircraft skin (0.020 in. thick) coated with 0.002 in. of standard white aircraft paint		Blisters	15	30	40
Cotton canvas sandbags, dry filled		Failure	10	18	32
Coral sand		Explodes (popcorning)	15	27	47
Siliceous sand		Explodes (popcorning)	11	19	35

[a]Radiant exposures for the indicated responses (except values marked with a superscript b, see footnote b) are estimated to be valid to ±25 percent under standard laboratory conditions. Under typical field conditions, the values are estimated to be valid within ±50 percent with a greater likelihood of higher rather than lower values.
[b]Ignition levels are estimated to be valid within ±50 percent under laboratory conditions and within ±100 percent under field conditions.
[c]Data not available or appropriate scaling not known.
[d]Radiant exposures for ignition of these substances are highly dependent on the moisture content.

surface detonation is more effectively shadowed by buildings, terrain, and other obstructions than is the light from an air burst (Miller 1962). The crater ejecta may also cover nearby fuel and smother incipient fires. In a subsurface explosion (where an armored penetrating warhead is used) the thermal pulse is substantially attenuated (GD77). Moreover, the base surge (caused by ejected material falling back upon the crater) could snuff out small fires and cover the fuel near the explosion site. Nevertheless, in a surface burst, it is still likely that primary thermal (and in cities, secondary blast-induced) fires would occur out to the ~2-psi overpressure contour (i.e., over an area of about 150 km^2 for a 1-Mt detonation; GD77). In buried explosions the situation is more complicated because both ground shock and air blast could contribute to secondary fire ignitions in cities. In any case, the present baseline scenario specifies air bursts against all urban and industrial targets, with only 30 percent (1500 Mt) of the remaining bursts detonated on the surface.

The fire effects of multiple nuclear detonations over cities and forests are complex and undetermined. Smoke from the fires of initial bursts could block subsequent thermal flash effects in some cases. Delayed bursts would probably spread existing fires, however, particularly by generating strong surface winds and convective plume activity. Closely spaced explosions over forests could greatly enhance the probability of fire ignition and spread. The problem of multiburst phenomena has not yet been adequately treated in the nuclear effects literature.

Urban Ignition

Some evidence that nuclear explosions are unique in their ability to ignite mass fires is offered by the Hiroshima and Nagasaki experiences. One crude estimate of the average energy release rate places the Hiroshima fire among the least intense of the mass fires of World War II (Martin, 1974). Nevertheless, centripetal winds characteristic of a firestorm apparently developed, and the fuel consumption within the fire zone was nearly complete (GD77; Ishikawa and Swain, 1981).

Some of the factors that affect nuclear fire genesis in cities are summarized in Table 5.3. Even though the blast wave that follows the thermal pulse could extinguish many of the primary thermal radiation fires, a substantial number of these ignitions would continue to burn. Idealized field tests to determine the efficiency of fire extinction by pressure waves are contradictory, and often little or no effect is observed (Wiersma and Martin, 1973; OTA, 1979; Backovsky et al., 1982). In fact, in one study, the blast dispersal of burning curtain fragments through a room was a major factor in fire development (Goodale, 1971). In addition, the blast ignites many secondary fires and creates conditions (Table 5.3) that strongly favor the growth and spread of the surviving fires. Overall, blast would appear to encourage mass fire development. The evidence from Hiroshima and

TABLE 5.3 Nuclear Mass Fires

Factors Contributing to Mass Fires

Thermal irradiation ignites materials at fluences of 10 to 20 cal/cm^2 over a large area.

Blast starts secondary fires out to ~2 psi overpressure.

Fires are started simultaneously over a large area.

Fires are ignited on both sides of major firebreaks.

Blast scatters solid fuels, ruptures gas and liquid fuel lines, opens windows, and breaches firebreaks.

Blast breaks water mains, blocks streets, and causes injuries that prevent effective firefighting.

Nuclear fireball rise establishes central convective motion of a mass fire.

Natural wind vorticity contributes to the formation of fire whirls and firestorms.

Factors Inhibiting Mass Fires

Blast wave extinguishes many fires started by thermal radiation.

Blast covers flammable materials with nonflammable debris.

Reduced visibility, due to natural or nuclear causes, limits thermal radiation effects.

Meteorological factors such as winds, high humidity, and precipitation retard fire spread and coalescence.

Nagasaki suggests that both primary and secondary fires eventually contributed to the conflagrations.

Detailed models of nuclear fire initiation and spread in urban and suburban settings have been constructed (Miller et al., 1970; Martin, 1974; FEMA, 1982), although their fidelity is in some doubt (Miller et al., 1970). The models suggest that, within the 20-cal/cm^2 irradiation perimeter, $\geq$20 percent of the buildings could have one or more initial fires. This assumes that the blast wave extinguishes almost all of the primary fires and, overall, inhibits fire growth and spread (FEMA, 1982). However, even if the initial fires are sparsely distributed after a nuclear explosion, nearly all blocks of houses or buildings are likely to have at least one fire (Martin, 1974). By implication, few effective firebreaks would exist in the initial fire zone. Observations of everyday urban fires indicate that fire spread between buildings (mainly by heat radiation and firebrands) is very efficient (~50 percent probability) at separations of about 7 m or less, and can occur over distances of 15 to 30 m (Chandler et al., 1963; Ayers, 1965; FEMA, 1982). Rows of residential homes, and certainly buildings in city blocks, are generally separated by less than 10 m. Accordingly, there is a high probability that 50 percent or more of these buildings would eventually burn out (Martin, 1974; FEMA, 1982). Owing to the dispersal of fuel by the blast into the gaps between the buildings, and the strong winds generated by the explosions and conflagrations, fire spread could be even more efficient in the nuclear case. Large isolated (industrial) buildings would also have a high probability of burning because of their large total area of exposure and therefore high likelihood of having at least one initial fire (Martin, 1974).

At blast overpressures of $\gtrsim$15 psi, concrete and steel buildings suffer severe damage and break apart to produce rubble. The area of such damage is about 25 km^2/Mt (GD77). In densely built up areas, the rubble could be several meters deep. Fires can burn in rubble, but generally at a slower rate. Obviously, civil defense and firefighting efforts would be futile under such conditions, and fire spread would be uninhibited by gaps and open areas. The buried fuels would tend to smolder and pyrolize in the heated air that filtered through the rubble, thus smoking copiously. It is expected that a large fraction of the combustibles in the rubblized zone would eventually burn, possibly with an exaggerated smoke emission confined to lower altitudes.

If an effective firefighting effort could be mounted, many of the initial urban fires might be extinguished and fire spread substantially limited in the lower overpressure regions (Kanury, 1976; FEMA, 1982). Such an expectation is probably optimistic. In Hiroshima and Nagasaki, even under wartime preparedness, firefighting efforts were largely futile (Ishikawa and Swain, 1981). Once the initial fires had grown to even moderate size, attempts at containment were hopeless without sufficient water, tools, and manpower. It follows that, within 1 to 2 h after a nuclear explosion over a city, major fires would be burning throughout the original fire ignition zone.

Forest Ignition

Little information is available on forest, brush, and grass fires initiated by nuclear explosions (Jaycor, 1980). Some factors that would influence the extent of nuclear wildfires are as follows:

1. The number of low air bursts over areas of forest, brush, and grass.
2. Meteorological conditions, such as cloudiness, precipitation, winds, humidity, and snow cover.

3. The probability of igniting persistent fires in the fuel bed, accounting for the shading of dry fuels by the live canopy.

4. The probability of fire spread in the fuel bed.

5. The effects of blast on the distribution of fuels and the development of fires.

6. Other factors, such as terrain, existence of firebreaks, and nearby nuclear explosions.

Rough estimates for some of these factors, based on past wildfire experience and theoretical analyses of nuclear effects, are discussed below.

Although Ayers (1965) had pointed out that many fires are likely to occur in a nuclear exchange, Crutzen and Birks (1982) made the first quantitative estimate of forest fire smoke and gas emissions in a nuclear war, and proposed that large quantities might be generated. As in cities, the nuclear bomb light is likely to ignite numerous small fires over a large area, most of which would be extinguished by the blast wave (Jaycor, 1980). The area initially subject to ignition could be as large as 500 km^2/Mt (Ayers, 1965), which corresponds to thermal fluences of $\gtrsim$10 cal/cm^2. It is possible that the number of individual fires surviving the blast wave and developing into major conflagrations could well exceed one per 10,000 m^2 (i.e., 100 ignitions per square kilometer). The rise of the nuclear fireball would establish strong afterwinds to fan the fires. It is unlikely that organized firefighting crews with sophisticated equipment would be available to extinguish the flames.

Nuclear forest fires would not resemble most forest fires of the past. It is conceivable, although uncertain, that, because of the simultaneous ignition over a large area and the fanning action of the afterwinds, some of the nuclear forest fires could develop into intense firestorms with towering smoke plumes. The distribution and consumption of fuel in nuclear forest fires could also be significantly modified. For one thing, much of the forest canopy and some heavy timbers would be shattered and blown down into the burning zone. If the nuclear fire were very intense, even large standing timbers could be substantially charred. Thus nuclear forest fires might consume a larger fraction of the forest fuels than typical natural wildfires (see below).

Huschke (1966) surveyed wildland fuel patterns and flammabilities in the United States as a function of region and time of the year. Some of his data are summarized in Figure 5.2. For summer conditions, up to 50 percent of all fuels (grass, brush, and timber) can have medium to high flammability at any given time (and presumably would be capable of sustaining type II nuclear fires). The total area involved is about 2 x 10^6 km^2, or about 30 percent of the continental United States. In winter, only about 10 to 20 percent of the fuels would exist simultaneously in such a flammable condition. Huschke's analysis is consistent with data on the geographical and temporal occurrence of "actionable" and "critical" fire conditions developed by Schroeder and Chandler (1966).

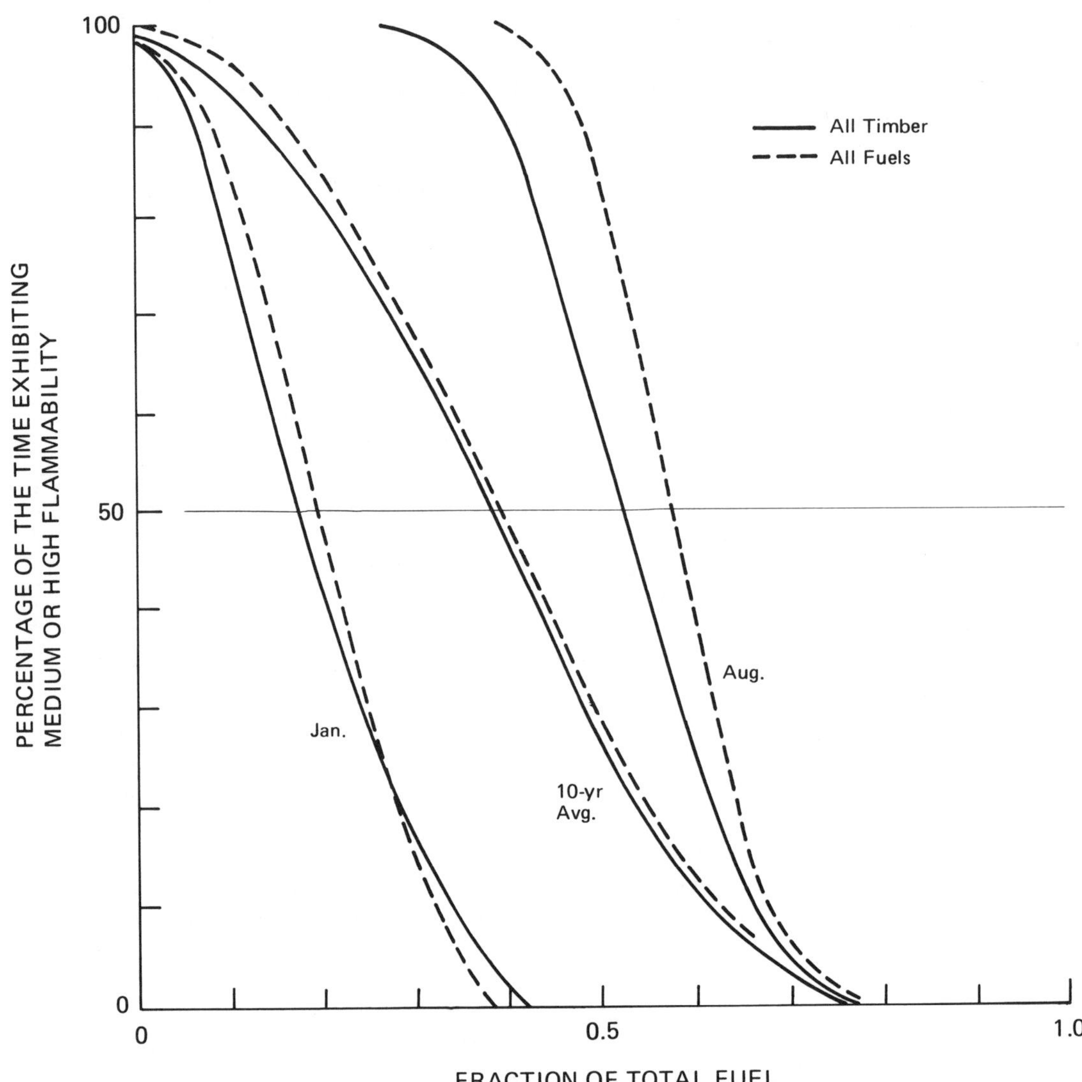

FIGURE 5.2 Flammability statistics for wildland fuels in the United States. The curves show the percentage of the time that various cumulative fractions of the fuel are in a state of moderate to high flammability. (From Huschke, 1966)

Accurate flammability statistics are not available for the Soviet Union, but similar fire conditions are likely to prevail at the same times of the year (e.g., Shostakovitch, 1925).

The potential for fire spread in wildlands is illustrated by U.S. Forest Service estimates of burnout areas for 1-Mt explosions, which range from 500 to 20,000 km^2 (Hill, 1961). Typical burnout areas would, of course, lie near the lower end of the range.

BURDENS AND DISTRIBUTIONS OF COMBUSTIBLE MATERIALS

By using published information, the quantity and distribution of combustible materials in cities and forests may be determined with reasonable accuracy for the purposes of this assessment.

Urban Combustibles

Surveys of combustible materials in urban settings are available (e.g., U.S. Department of Defense, 1973; Culver, 1976; Issen, 1980; FEMA, 1982). Some of the data are summarized in Table 5.4. Flammables in urban zones include leaves and brush, wood framing and siding, paper and cardboard, stored gasoline and oil, tar and asphalt, rubber, natural and synthetic fabrics, plastics, paints, solvents, and a variety of other household and industrial chemicals. Worldwide anthropogenic production of combustible materials includes about 2000 Tg/yr of timber (roughly half of which is burned as fuel), about 2000 Tg/yr of liquid fossil fuels, about 100 Tg/yr of tar and asphalt, about 150 Tg/yr of oil-based lubricants and solvents, about 50 Tg/yr of plastics and resins, about 30 Tg/yr of natural and synthetic fibers, and about 10 Tg/yr of rubber (U.N., 1981). These materials have been accumulating in cities for up to 50 years. At any time, perhaps 1000 to 2000 Tg of liquid and gaseous fossil fuels and related compounds are stored in urban areas. In city centers, combustible burdens can exceed 100 g/cm^2. However, over the great expanses of modern urban and suburban complexes, and in smaller cities and towns, average combustible loads of 1 to 5 g/cm^2 are more likely (FEMA, 1982; Turco et al., 1983b; Crutzen et al., 1984). An average urban combustible burden of 4 g/cm^2 is assumed in the present baseline case to represent the mean distribution between the city centers and the suburbs in metropolitan complexes (although excursions to lower burdens are considered below).

Studies of urban and suburban areas worldwide indicate about 2300 cities with populations exceeding 100,000. The corresponding urbanized area is roughly 1.5 x 10^6 km^2. There are about 180 cities with populations exceeding 1,000,000 (Turco et al., 1983b, and references therein). About 85 percent of the world's urban areas are located in the northern hemisphere. In the NATO and Warsaw Pact countries (which includes the United States, USSR, and most of Europe), there are ≈1100 cities with populations greater than 100,000 and ≈80 major urban areas with populations exceeding 1,000,000 (the major urban agglomerations include some of the smaller cities). The total urban and suburban area involved is about 500,000 km^2, and the co-located population is about 500,000,000. In the United States, cities tend to sprawl over large areas, while in the USSR and Europe, the cities are more compact. The total area of the "core" cities in the NATO and Warsaw Pact countries is roughly 10 percent of the total urbanized area, or about 50,000 km^2. These central urban zones, in which industrial/economic targets are concentrated, and near which significant military targets are often located, are likely to be hit by

TABLE 5.4 Typical Combustible Burdens in Urban Areas

Building Type	Average Combustible Burden per Story[a] (g/cm^2)	Number of Stories	Building Density[a] (percent)	Overall Combustible Load (g/cm^2)
Residential:				
brick or frame	5-10	1-2	10-25	0.5-5
Office and commercial	5-20	2-10	20-40	2-80
Industrial	0-15	1-3	20-40	0-18
Storage	10-40	1-2	20-40	2-32

[a]Data taken from FEMA (1982).

nuclear explosions, are heavily loaded with combustibles, and are expected to burn vigorously.

Considering the typical design of cities, with most facilities and activities concentrated at the center, it appears that less than 10 percent of the total urban area may hold 50 percent or more of the total urban combustible material. The urban centers of the world may hold a total of 10,000 Tg or more of flammables, while the surrounding suburban areas may contain an equal amount. It is estimated that up to three-quarters of all the combustible material consists of wood and wood products, with perhaps 5 to 10 percent in plastics, resins, and rubbers, and the remainder in oil, tar, asphalt, gasoline, solvents, and other fuels and organochemicals. At the present time, no comprehensive inventory of urban combustible materials is available.

Crutzen and Birks (1982) pointed out that oil and natural gas production and storage fields might be targets of nuclear explosions. Fires in some uncapped wells, once ignited, could burn uncontrolled for months or longer. Crutzen and Birks estimated that the rate of fuel release and combustion might equal the current world production rate, or about 300 Tg per month. Although such fires are potentially important if ignited, they are not included in the baseline analysis.

In some very intense urban fires, materials not usually thought of as "flammable" could ignite and burn, or decompose into vapors that later nucleate into particles. Of particular interest is aluminum, which is used in large quantities in modern construction, which burns readily, and which forms about 2 g of smoke per gram of aluminum burned. Other compounds containing magnesium, calcium, and sodium could generate additional inorganic smoke under extreme conditions. These possible inorganic sources of smoke are omitted from the baseline case because of the difficulty in estimating the quantities involved.

Forest and Wildland Combustibles

Forests cover about 40 percent of the land surface of earth, or $\approx 4 \times 10^7$ km^2, and brush, grass, and pasture lands account for perhaps another $\approx 4 \times 10^7$ km^2. In the United States, Canada, Europe, and the USSR, the total forested area is $\approx 1.8 \times 10^7$ km^2, or about 40 percent of the total land area. The average concentration of combustible biomass in temperate and boreal forests is ≈ 2 g/cm^2 (Seiler and Crutzen, 1980; USDA, 1981). In natural wildfires, up to 25 percent of this fuel is burned, consisting of organic compost, ground litter, understory growth, leaf canopy, and small branches and twigs (Wright and Bailey, 1982). Because of the potentially unique character of nuclear forest fires, a larger quantity of fuel might be consumed, as noted earlier. Nevertheless, it is assumed here that 0.4 g/cm^2 of fuel would be burned in nuclear forest fires, which is typical of many natural forest fires (Wright and Bailey, 1982; Crutzen et al., 1984).

Brush and grasslands have much lower fuel burdens than forests ($\lesssim 1$ g/cm^2), but more of the fuel ($\gtrsim 50$ percent) is likely to burn. Accordingly, fuel consumption in nuclear brush fires might be estimated as about 0.1 to 0.2 g/cm^2, and in nuclear grass fires, about 0.02 to 0.05 g/cm^2 (USDA, 1972; Wright and Bailey, 1982). Smoldering fires in peat soils have been known to burn out accumulated organic layers several feet thick (Wein and MacLean, 1983). Nevertheless, brush, grass, and peat fires are ignored here.

Urban Combustibles Consumed

In a full-scale nuclear exchange, attacks against military, industrial, and economic targets in and around cities might involve multiple detonations to achieve complete destruction. This could limit the average effective area and combustible load ignited per megaton of explosive yield. A number of other complex factors could further limit the quantity of combustible materials exposed to nuclear fire, while some factors could increase the quantity. Factors that could limit the potential urban fire area and/or the burden of flammables consumed include the following:

1. Direct overlap of thermal irradiation zones when detonations are closely spaced.
2. Multiple targeting (two or more consecutive explosions on one target) in order to assure destruction.
3. Obscuration of bomb light by smoke and dust generated by previous explosions and fires.
4. Regions of high overpressure (8 to 15 psi), which produce more rubble and bury more flammable material.

5. Emphasis on "counterforce" (military) targeting in many strategies.

6. Geographical factors (bodies of water) and topographical factors (hills) that delimit or shield possible ignition areas.

7. Civil defense efforts.

Factors that could increase the potential fire damage include the following:

1. Confinement of the heaviest combustible loadings within the areas most likely to suffer multiple detonations, i.e., co-location of flammable materials and key urban and industrial targets.

2. Attack strategies that employ damage assessment--particularly in the case of large, immobile, and highly vulnerable urban and industrial targets--to conserve and optimize the use of nuclear forces.

3. Enhancement of fire ignition and spread probabilities for multiple bursts over targets, due to heating by additional thermal irradiation, fanning by surface and convective winds, and spreading by firebrands.

4. Fire ignition and spread well beyond the 4- to 5-psi region, as was observed during World War II.

5. Collateral damage to urbanized areas from attacks against military targets, for which there is (a) less inherent overlap of the fire zones and (b) ignition of the perimeter of cities with increased likelihood of spread into the city centers (Larson and Small, 1982a)

6. Localized enhancements in nuclear thermal irradiation due to reflection from clouds.

7. Fire damage from tactical nuclear and conventional weaponry.

Obviously, a detailed calculation of the urban ignition area in a nuclear conflict would require an enormous amount of information, much of which is not easily obtainable. For the present assessment, certain simplifying assumptions have therefore been made. Of the 1500 Mt detonated over urban areas in the baseline scenario (see Chapter 3), 500 Mt is assumed to be ineffective due to overlap of thermal irradiation zones. In accordance with previous discussions, the average fire area per megaton of yield is taken to be 250 km^2/Mt for the other 1000 Mt. The industrial, economic, and co-located military targets in this 1000-Mt attack are assumed to be distributed among the approximately 1000 largest cities of the NATO and Warsaw Pact countries (Chapter 3); these cities collectively have an urban/suburban area of about 500,000 km^2. Thus, by implication, fire ignition would occur over 50 percent (250,000 km^2) of the developed area of the warring nations. While military facilities in China, Japan, and other countries might also be targeted, nearby urban areas are not assumed to be affected.

The average combustible burden is assigned a value of 4 g/cm^2 for urban/suburban construction. That is, regions of high combustible loading (for example, central cities, where flammable burdens can reach 100 g/cm^2 or more) are averaged together with regions of low loading to account for the likely fire damage to the vast suburban and residential areas of the developed nations. It follows that 10,000 Tg of combustible material would be subject to nuclear ignition in the baseline case.

The question of overlap in urban fire zones is probably not critical to this analysis because (a) anthropogenic flammable materials are highly concentrated in relatively small areas (in cities), and (b) urban nuclear attacks would be likely to trigger conflagrations that would spread outside of the original ignition zones. For example, if the present 1500-Mt urban attack were assumed to be concentrated over the city centers, with an areal ignition overlap factor of 10 (i.e., with an average ignition area of only 25 km^2/Mt), 37,500 km^2 of the central cities with the densest combustible loading could still burn. For a central city fuel burden of 20 g/cm^2, nearly the same total quantity of combustible material would be impacted as in the baseline case. The inevitable spread of urban conflagrations would ensure even greater fuel consumption. Given that up to 1000 cities could be affected in the baseline exchange, such extensive overlap (a factor of 10) is very unlikely in the first place.

The low sensitivity of the quantity of urban fuel impacted to the total yield of an exchange is also demonstrated by the 100-Mt excursion scenario of Turco et al. (1983a). In this case, a 100-Mt attack with 100-kt warheads directed exclusively against the largest built-up city centers (and spaced to ignite about 25 km^2 per 100 kt) was found to consume roughly the same amount of combustible material as a generalized 1000-Mt urban/suburban attack.

Other approaches could be taken to estimate the areas burned and flammables consumed in urban nuclear fires. For example, a comprehensive target analysis might be carried out. This would require detailed information about the precise locations of key industrial and military facilities, the flammable materials at these locations and over large surrounding areas, and the plan of attack to disrupt production and military operations. Clearly, unless all potential targets and a number of attack strategies were considered, the estimation of impacted urban flammables might not be significantly improved. Hence, the simplified approach adopted in this report seems reasonable at this time, and is in quantitative agreement with estimates of material consumption obtained by a number of other schemes (Turco et al., 1983a,b; R. Turco, private communication, 1984; Broyles, 1984; Crutzen et al., 1984).

Forest and Wildland Fuels Consumed

As in the case of urban fire ignition, a number of complex factors could affect the area and the quantity of wildland fuels consumed in a nuclear exchange. Factors that could limit the potential wildfire area and fuel consumption include the following:

1. Overlap of target zones, particularly in missile silo fields.
2. Multiple bursts over isolated "hard" military targets.

3. Smaller ignition areas for very low air bursts and ground bursts aimed at "hard" targets.

4. Obscuration of bomb light by dust and smoke from previous explosions.

5. Restricted flammability of wildlands due to meteorological and seasonal effects.

6. Firebreaks and firefighting.

Factors that could augment wildland fuel consumption include the following:

1. Multiple explosions, which would increase the probability of fire ignition and spread in forests because (a) the blast would knock down the vegetative canopy, which is a major factor limiting fire ignition by single explosions over forests, (b) the repeated thermal irradiation and winds would dessicate fuels, augmenting fire ignition and spread probabilities, (c) the explosion winds would spread firebrands and fan established fires, and (d) natural clouds and the dust clouds created by ground bursts would scatter back and intensify the thermal radiation field of very low altitude detonations.

2. Location of a substantial fraction of the hundreds of military bases (other than missile fields) within several kilometers of forested areas; up to 2000 Mt could be directed against these targets.

3. Tactical nuclear warfare (in Europe) involving explosions over heavily forested areas (used for camouflage); tactical weapons are particularly effective in forest fire ignition (Woodie et al., 1983). About 500 Mt of tactical weapons are detonated in the baseline scenario.

4. Frequent spread of fires well outside of the ignition zone; occasionally, an individual fire might spread over 10,000 km^2 or more.

5. Consumption of vast quantities of other wildland fuels such as grass and brush.

6. On military bases, particularly strategic air bases, burning of large stores of aircraft fuels, buildings, munitions, and other materials.

It is assumed in the baseline case that 5000 Mt of explosions are detonated over widely dispersed military targets. Ignition probabilities are greatest in late spring, summer, and early fall (Schroeder and Chandler, 1966). Attacks at other times of the year would produce fewer fires. On a purely random basis, about 40 percent of the military explosions could occur over forests, with 40 or 50 percent of the remaining explosions occurring over brush- and grass-covered lands. Since, according to Huschke (1966), roughly 50 percent of such areas are capable of sustaining type II fires at any given moment under summertime conditions (Figure 5.2), an average total ignition yield of about 2000 Mt might be expected. Of course, many of these explosions would overlap, so that an effective ignition yield of about 1000 Mt may be more reasonable. The corresponding total ignition area would be 250,000 to 500,000 km^2, corresponding to a thermal fluence of 10 to 20 cal/cm^2 (e.g., Hill, 1961; Ayers, 1965).

Missile silo fields occupy an area of about 250,000 km^2, counting fringe regions. The silos are spread out to reduce vulnerability. Based on a survey using Landsat photographs (Short et al., 1976), and coarse vegetation maps, about 20 percent of the land housing silos appears to be forested (i.e., 50,000 km^2). A 2000-Mt overlapping barrage against these silos would probably incinerate these regions almost entirely.

Other military attacks would account for about 3000 Mt of additional explosions (after deducting the silo yield of 2000 Mt and the urban yield of 1500 Mt from the total baseline yield of 6500 Mt). If 40 percent of these explosions occurred over forests and ignited 250 km^2/Mt half of the time, another 150,000 km^2 of forest land could burn. This assumes that areas affected by only one explosion at irradiation levels of 20 cal/cm^2 or greater have a 50 percent probability of ignition (Huschke, 1966) and that areas affected by two explosions, each at $\geq$20 cal/cm^2, have a much greater probability of burning because of the drying effect of nuclear thermal radiation on exposed vegetation.* It is further assumed that, in this category of targeting, more than double overlapping at such high thermal intensities and overpressures is unlikely. Finally, fire spread is assumed to increase the overall area of burnout by about one third, allowing as well for some ignitions outside of the 20 cal/cm^2 zone (Hill, 1961; Ayers, 1965).

Military barrage attacks, in which overlap is purposefully minimized to optimize the area of impact, are omitted. Such attacks could be directed at the strategic bombers scrambling from air bases, at mobile missiles hidden in multiple bunkers or dispersed in forests. It should be obvious that such tactics would significantly increase the areas of wildfires.

The total area of forest fires in the baseline nuclear war is thus taken to be 250,000 km^2, with all other related fires neglected. The total combustible fuel within this area is about 5000 Tg, of which 20 percent is assumed to burn (see the previous sections). An area of 250,000 km^2 seems reasonable, in view of the physical characteristics of the targets and the flammability statistics of wildlands. In previous generalized analyses, Crutzen and Birks (1982) estimated a forest fire area of 1,000,000 km^2; Turco et al. (1983a), 500,000 km^2; and Crutzen et al. (1984), 200,000 to 1,000,000 km^2. On the other hand, Small and Bush (1984), using a more detailed targeting and fire ignition methodology, have estimated a forest burn-off area of only 70,000 km^2 (and a total fire area of 170,000 km^2) in a summertime exchange of about 4000 Mt. The differences between the highest ($\approx 10^6$ km^2) and lowest ($\approx 10^5$ km^2) forest fire area estimates have yet to be resolved.

*The latter assumption must be tested, but is based on the noticeable "greying" of live vegetation by bomb light during nuclear tests.

SMOKE EMISSIONS AND PROPERTIES

The dense smoke plumes from massive urban and natural fires have been described by many observers. Downwind of large conflagrations there are often reports of dark skies, red suns, and black rain (e.g., Plummer, 1912; Lyman, 1918; Ishikawa and Swain, 1981). Photographs show black plumes rising over industrial fires, and satellite images show wildfire plumes extending downwind for hundreds of kilometers--direct evidence that large fires can cause profound local optical and physical perturbations of the atmosphere (e.g., Davies, 1959).

The important properties of smoke are the quantity generated per unit mass of fuel consumed, the particle composition and size distribution, the specific extinction and absorption coefficients (expressed in square meters per gram) at visible and infrared wavelengths, and the heights of injection into the atmosphere. All of these properties are highly variable, depending on fuel type, moisture, burning conditions, and so on.

Experimental evidence suggests that the smoke from a composite array of fuels in a fire may be roughly summed over the array (Rasbash and Pratt, 1979), although in certain instances the average smoking tendency of the mixture can be lower than the sum for the individual components (Rasbash and Drysdale, 1982). Using an approximate linear addition rule, average values for smoke characteristics can be deduced if the fuel array is adequately defined. In terms of the optical properties of the smoke, an attempt is made here to be somewhat conservative in minimizing the possible sooty, highly absorbing component.

Urban Smoke

Emission Factors*

Many of the flammable materials that are commonplace in homes, businesses, and industry can generate dense, sooty smoke and toxic gases when burned in an uncontrolled environment. The mechanisms of soot formation in flames (Calcote, 1981; Nakanishi et al., 1981) and smoldering fuels (Bankston et al., 1981) have been investigated, but no complete theory has yet been established. A variety of molecular organic neutral and ionic precursors in the flame may combine to form incipient soot (graphitic and amorphous elemental carbon) particles, which continue to grow by accretion of organic material while losing hydrogen relative to carbon. Several studies have been made of the soot produced by organic gas flames (Chippett and Gray, 1978; Pagni and

---

*The smoke emission factor can be expressed as the mass of smoke generated per unit mass of fuel burned--g-smoke/g-burned--or as a weight percentage yield of smoke--percent of burned mass converted to smoke. The factors are interchangeable; i.e., a 4 percent smoke yield by mass is equivalent to an emission factor of 0.04 g/g.

Bard, 1979; Kent and Wagner, 1982), anthropogenic liquid and gaseous fuel combustion (Day et al., 1979; Kittelson and Dolan, 1980; Wolff and Klimisch, 1982), and burning solid materials (Hilado and Machado, 1978; Ohlemiller et al., 1979; Jagoda et al., 1980; Bankston et al., 1981; Vervisch et al., 1981; Butcher and Ellenbecker, 1982; Tewarson, 1982). Some properties of smoke produced by burning wood and polymers are given in Table 5.5.

The quantity of smoke and soot generated by a fire depends on several factors, including the efficiency of ventilation of the fire and the average temperatures in the pyrolysis and burning zones (Quintiere, 1982). In controlled combustion devices, such as oxyacetylene torches, oil burners, and automobile engines, smoke and soot emissions are minimal (unless pretuned operating conditions are disturbed). Even a crackling fire in a fireplace represents well-controlled combustion (although particulate emissions can approach 1 to 2 percent (by weight) of the wood burned (Dasch, 1982)). In laboratory experiments on smoke generation, the smoke emission during flaming combustion is generally observed to increase rapidly as the oxygen available in the burning zone decreases (Saito, 1974; Morikawa, 1978; Tewarson et al., 1980; Tewarson and Steciak, 1982), or as the ventilating air supply is preheated (Morikawa, 1978; Powell et al., 1979; Bankston et al., 1981). The smoke output of flaming wood decreases as the sample is heated artificially by radiation (Bankston et al., 1981), although the opposite effect is seen in other materials (Seader and Einhorn, 1976; Tewarson, 1982). In nonflaming combustion, smoke emissions typically increase markedly when the samples are radiatively heated, but decrease in certain materials such as polyurethane foam when the air supply is preheated (Seader and Einhorn, 1976; Bankston et al., 1981).

In unregulated fires, the efficiency of ventilation by winds and turbulence should tend to decrease as the fire area and fuel density increase. Accordingly, with an expansion in the scale of a fire, the smoke emission factor should increase (Rasbash and Drysdale, 1982). This effect is generally, but not always, observed under experimental conditions (Quintiere, 1982; Rasbash and Drysdale, 1982). On the other hand, the intensification of a fire with size tends to reduce smoke emissions through several mechanisms: augmented winds and turbulence, which enhance local ventilation; higher temperatures, which incinerate the smoke over the fire; and induced precipitation, which scavenges the particles. The smoke from very intense fires, although reduced in quantity, is enriched in graphitic carbon (soot) (Morikawa, 1978; Wagner, 1981), and thus absorbs light more effectively (see below). The complete incineration of soot in flames requires temperatures in excess of 1500 K (Wagner, 1981), whereas the temperatures above flames, even in very intense fires, are typically <1000 K. Hence it is not clear that violent burning would significantly reduce, through reduced smoke emissions, the long-term optical effects of large-scale fires.

The complexity in predicting the smoke-forming properties of materials can be illustrated by using laboratory test data for construction lumber. During flaming combustion fully ventilated by room-temperature air, Bankston et al. (1981) found that Douglas fir

TABLE 5.5 Examples of Measured Smoke Properties[a]

Material	Flaming Conditions ε (%)	Flaming Conditions r_M (μm)	Flaming Conditions f_c	Smoldering Conditions ε (%)	Smoldering Conditions r_M (μm)	Smoldering Conditions f_c
Polyvinylchloride	1.2-2.6	0.2-0.5	0.2	7.2-55.	0.3-0.5	--
Polyurethane foam	--	--	0.02-0.10	5.7-37.	0.25-0.45	0.01-0.09
Polystyrene	~3.2	~0.6	~0.2	8-20	~0.7	--
Polypropylene	~1.6	~0.6	~0.1	~12.	~0.8	--
PMMA	<1	~0.6	~0.04	0.5-5.6	~0.3	--
Polyethylenes	--	--	~0.06	8-40	--	~0.10
Epoxy fiberglass	--	--	~0.06	12-27	--	--
Plastics[b]	6-20	--	--	--	--	--
Plastics	3-5		>0.9[c]			
Rubber	~10	--	>0.9[c]	--	--	--
Oil slick	>2-6	--	>0.9[d]	--	--	--
Construction lumber	0-2.5	~0.2	0.08-0.3	3-16.5	0.17-0.6	--
Pine needles	0.5-1.3[e]	--	~0.15	1-6[f]	--	~0.03
Forest fires[g]	1-7	~0.05	0.03-0.15	--	--	--

[a]Tabulated smoke properties are as follows: ε, the weight percent of burned material converted to smoke particles; r_M, the mass median particle radius; f_c, the mass fraction of graphitic carbon in the smoke. Ranges of values partly reflect differences in burning rates and ventilation. Mass median radius, r_M, is ≃4.3 times larger than number mean (mode) radius for a log normal size distribution with $\sigma = 2$.
[b]The relative importance of flaming and smoldering conditions was not described.
[c]Soot yield, which includes an unknown fraction of oils (presumably <10 percent).
[d]On the basis of microscopic analysis, it is assumed that the soot is nearly pure elemental carbon.
[e]Refers to "backing" fires (fires moving against the wind).
[f]Refers to "heading" fires (fires moving with the wind).
[g]Includes a combination of flaming and smoldering conditions.

SOURCE: Data are taken from Hilado and Machado (1978), Bankston et al. (1981), Tewarson (1982), Patterson and McMahon (1983), and Crutzen et al. (1984).

irradiated by a supplementary heat source of 2.5 W/cm^2 had a smoke emission factor of about 2.5 percent. At an irradiation of 5 W/cm^2, however, the emission factor was <1 percent. In other flaming tests with Douglas fir at irradiation levels of 5 W/cm^2, Powell et al. (1979) observed a steady increase in smoke production as the air flow was heated from about 300 K to 600 K. By contrast, with smoldering Douglas fir, a substantial decrease in smoke production (from a reference yield of about 15 percent) was measured as the ventilating air was heated from 300 K to 600 K (Powell et al., 1979). Saito (1974) noted that dry spruce and other wood products generated less smoke as flaming samples were heated from about 750 K to 850 K. In larger-scale chamber tests, however, 2 to 4 times as much smoke was produced at the higher temperatures as in the individual sample tests (Saito, 1974). Finally, room-size experimental wood fires were found to emit, on the average, about 3 to 6 percent smoke by weight (Rasbash and Pratt, 1979).

The smoke emissions from a variety of noncellulosic urban combustibles (e.g., oil, tar, plastics, and rubber) range from about 5 to 40 percent (of the weight of fuel consumed) under smoldering conditions to 1 to 15 percent in flames (Table 5.5). In large building and area fires, all burning conditions exist simultaneously, and an average smoke emission factor of 6 percent is probably reasonable for these materials (Seador and Einhorn, 1976; Rasbash and Pratt, 1979). For dry construction lumber, furniture, paper, and other cellulosic materials, an average flaming/smouldering smoke emission factor of 3 percent can be used (Rabash and Pratt, 1979).

In subsequent baseline estimates, an average smoke emission factor of 4 percent is adopted for all urban fires (<u>before</u> smoke scavenging and removal in the fire plumes--see below). This average emission figure assumes that two-thirds of the material burned is cellulosic with an emission factor of 3 percent, and one-third is noncellulosic or polymeric with an emission factor of 6 percent. For urban targets--industrial complexes and fuel production, distribution, and storage centers--the latter class of combustibles is even more common, and thus the relatively high percentage of these materials in the burned inventory. In older cities and residential areas, wood construction dominates the flammable burden.

Flame retardants applied to certain materials, such as flexible polyurethane foams, may actually increase their smoke output (Bankston et al., 1981). Most recently, smoke retardants have been developed for commercial applications, but these are not yet in wide use.

Size Distribution and Composition

The size distribution of smoke particles shows a significant variability (Table 5.5). Most sooty smokes are composed of nearly spherical ultrafine carbon grains of about 0.005 to 0.02 μm in radius and varying amounts of oils and tars. In dry smokes the carbon grains coagulate into chain structures with dimensions of the order of 0.1 μm. In oily smokes, droplets of heavy organic liquids laced with carbon grains can grow to sizes exceeding 0.2-μm radius. In either case, individual smoke particles may be composed of 10 to 1000 or more carbon grains. At very high rates of pyrolysis, carbon particles often agglomerate to sizes of 0.5 μm or larger (Bankston et al., 1981). In thick carbonaceous smokes, loose clusters of soot can grow to 10 μm, with occasional fluffy agglomerations reaching 100 μm (Day et al., 1979). (However, in these extreme cases it is suspected that individual soot clusters may aggregate on the filters used for collection.)

The elemental (graphitic) carbon fraction of the smoke generated during the unregulated combustion of liquid fossil fuels and synthetic fibers and plastics can range from 5 to 90 percent by weight (e.g., Tewarson, 1982), but is typically greater than 20 percent (Table 5.5). Burning wood generally produces less graphitic carbon (~20 percent). As was noted in the previous section, smoke from very hot, intense fires has a higher graphite content independent of the fuel. Considering the types of combustibles and burning conditions expected in urban nuclear fires, an average graphitic (soot) fraction of about 20 percent for the smoke seems conservative. Urban smoke generally absorbs sunlight more effectively than forest fire smoke, which is typically about 10 percent graphite (see below).

The baseline urban smoke emission factor of 4 percent (0.04 g/g) implies an elemental carbon emission factor of 0.8 percent. Accordingly, in the nominal case, the generation of light-absorbing particles in city fires is taken to be quite <u>inefficient</u>.

The size distribution of smoke particles may be represented by a log normal distribution,

$$n(r) = \frac{n_0}{(2\pi)^{1/2} \ln \gamma} \frac{1}{r} \exp \left[- \frac{\ln^2 (r/r_m)}{2 \ln^2 \gamma} \right] ,$$

where $n(r)dr$ is the number of particles in the size interval $r \rightarrow r + dr$ per cubic centimeter of air (i.e., n has dimensions of particles per cubic centimeter of air per unit radius interval), r_m is the number median (or mode)* radius, and γ is a measure of the dispersion or width of the size distribution. Here, the radius is the spherical volume-equivalent radius for odd-shaped particles. The density, ρ, of pure graphitic carbon is ≈ 2.25 g/cm^3; for most oils, $\rho \lesssim 1$ g/cm^3. Smoke particles that are composed predominantly of oils will be spherical and have a density of ≈ 1 g/cm^3. Smoke particles that are nearly pure elemental carbon will be characterized by a fluffy chained structure with a low effective ratio of mass to size (i.e., an effective density of $\ll 2.25$ g/cm^3).

---

*For the log normal size distribution chosen, r_m is both the number median radius (above and below which lies exactly half of the particles) and the mode radius (at which lies the peak in the number distribution plotted in ln r coordinates).

The value of γ for a wide range of natural and man-made aerosols falls in the range of 1.5 to 2.5. A reasonable average choice for smoke is γ = 2.0, which is consistent with a number of measurements (Turco et al., 1983b). In the log normal size distribution with γ = 2, the mass median particle radius* is approximately 4.3 times the number median radius.

Several factors may lead to changes in the size distribution of smoke particles within the plume of a mass urban fire:

1. Coagulation of the particles at high concentrations.
2. Collection of the particles by water droplets followed by coalescence of the smoke as the droplets evaporate.
3. Scavenging of the fine smoke aerosols on the surfaces of larger ash and debris particles.

Coagulation. Coagulation by Brownian diffusion is undoubtedly responsible for many of the large soot particles observed in thick smoke plumes (Day et al., 1979). (Aggregation of particles on filters and other sampling surfaces is responsible for occasional misinterpretations, however.) Because of coagulation, smoke concentrations in a plume are not expected to exceed approximately $10^5/cm^3$ after several hours, or approximately $10^4/cm^3$ after several days (Twomey, 1977). Turco et al. (1983a,b) carried out detailed coagulation simulations for both confined and dispersed smoke clouds, in which the evolving size distribution was predicted as a function of time and height. These size distributions were then used to perform optical calculations, assuming that the particles are always spherical. Turco et al. showed that, in artificially confined plumes with slow dispersion, the smoke particles could coagulate to average sizes of 0.5-μm radius, and that their optical depth, when related to the equivalent value for smoke instantaneously distributed over the northern hemisphere, was reduced by about a factor of 2 after about 1 month. This relatively small difference in optical depths followed because (a) the initial smoke clouds at different altitudes were not equally dense and so the particles coagulated to different extents, and (b) the size distribution changed in dispersion (γ) as well as in average size (r_m), which reduced the overall impact on the optical properties. Such results indicate that the effects of coagulation on the optical characteristics of smoke are not straightforward. Turco et al. also showed that, even if the fire plumes are instantly dispersed over the hemisphere, the smoke particles eventually coagulate to sizes up to about 0.2 μm (after about 1 month). Crutzen et al. (1984) analyzed the coagulation problem using "self-preserving" size distributions. However, for the present analysis, an even simpler approach is adopted which considers only the mean particle size.

---

*This is the radius above which, and below which, half of the total particle mass lies.

A simple model of an expanding fire plume may be used to derive the following approximate expression for the change in the average particle radius with time due to coagulation:

$$r/r_0 = \left[1 + \frac{1}{2} K n_0 \frac{\ln(1 + \alpha t)}{\alpha} \right]^{1/3} .$$

The initial particle radius is r_0, and the initial particle concentration is n_0 (particles per cubic centimeter). K is the Brownian coagulation kernel (~1 x 10^{-9} cm^3/s for particles of about 0.1 μm in radius), and α is the linear plume expansion coefficient (per second); i.e., α^{-1} is the time in which the initial plume volume doubles. Table 5.6 illustrates the dependence of the particle size on the parameters n_0, α, and t. It can be seen that, except in the most extreme cases of high initial particle concentration and retarded plume dispersion, the particles undergo only modest growth by coagulation in the first day after emission. This treatment assumes that Brownian coagulation is the dominant aggregation mechanism, and that the effects of electrical charge are small. A normal Boltzmann distribution of charge on the aerosols has little influence on the coagulation rate (Twomey, 1977).* Turbulence in fire plumes, which might be comparable in intensity to the turbulence in natural convective systems, is also a secondary factor (Rosenkilde and Serduke, 1983).

Baum and Mulholland (1979) simulated smoke coagulation in a buoyant plume using a coupled dynamical/microphysical model. They found that coagulation was significant only with very large initial particle concentrations, exceeding ~$10^8/cm^3$. In such cases, the coagulation process still tended to "freeze out," or turn off, rapidly. They estimated that particle radii could double under such extreme conditions. The data suggest that the dilution of smoke in a typical buoyant plume occurs within minutes of emission, which may not necessarily be the case in a very large plume. Tsang and Brock (1982) describe the effect of coagulation on the optical extinction of extended aerosol plumes propagating through the atmosphere; they show that the effect is quite small even for very strong sources of 0.1-μm particles.

Estimates of initial smoke particle concentrations just above a fire can be obtained from experimental data, and used to place realistic limits on n_0 for open, uncontrolled combustion. Palmer (1976) directly observed smoke particle concentrations of about 4 x $10^5/cm^3$ (with an average radius of about 0.1 μm) just above a large, intense test fire in the Flambeau series. Concentrations of smoke particles measured in prescribed forest fire plumes are typically about $10^5/cm^3$ in the plume core close to the fire (Packham and Vines, 1978). Benech (1976) determined the air volume flow rate

---

*In fact, electrical charge may inhibit the agglomeration of soot particles (Valioulis and List, 1984).

TABLE 5.6 Relative Sizes[a] of Coagulating Particles in an Expanding Smoke Plume

n_0 (particles/ cm^3)	Coagulation Time = 1 hour α (s^{-1})			Coagulation Time = 1 day α (s^{-1})		
	10^{-3}	10^{-4}	10^{-5}	10^{-3}	10^{-4}	10^{-5}
1×10^6	1.2	1.3	1.4	1.5	2.3	3.2
1×10^5	1.02	1.05	1.06	1.07	1.3	1.6
1×10^4	1.00	1.00	1.01	1.01	1.04	1.1

[a]The ratio of the final particle radius to the initial radius is given for each set of physical parameters. The particles are assumed to retain a spherical shape.

through a 600-MW fire at the Meteotron facility, which burns fuel oil. Since the rate of energy release in a fire can be related to the rate of fuel consumption and thus to the rate of smoke production, the calculated rate of air mass flow through the fire may be used to deduce an initial smoke particle concentration. Assuming a smoke emission factor of 0.04 g/g and a heat of combustion of 4×10^4 J/g for oil, Benech's data imply a maximum concentration of about $1 \times 10^6/cm^3$ of 0.1-μm smoke particles just above the fire. Carrier et al. (1984) simulated the plume dynamics of a very large urban mass fire. Their results suggest a peak smoke concentration of about $2 \times 10^6/cm^3$ of 0.1-μm mode radius particles. The calculations also indicate that the plume expands in volume by a factor of 10 in the first ~100 s.

Observations and simulations of fire plumes suggest that maximum initial smoke particle concentrations (n_0) are about $10^6/cm^3$. The initial expansion and dilution rate of the plumes are also very rapid, with $\alpha >> 10^{-3}/s$ (e.g., Benech, 1976). Therefore, according to Table 5.6 prompt coagulation should be rather limited in most plumes, unless the fires are unusually smoky or the plumes unusually compact. After a period of $\lesssim$1 hour, the maximum smoke concentration in a plume would probably be about $10^5/cm^3$, due primarily to dilution by the entrainment of ambient air.

In a series of measurements of the sooty aerosols generated by the oil-burning Meteotron facility, Radke et al. (1980a) and Benech et al. (1980) recorded the process of smoke aging through coagulation. Initially (in the rising plume), the smoke was dominated by very small particles (<0.05-μm radius). After 18 min, the concentration of these particles had begun to decrease in relation to the concentration of particles of >0.05 μm. By 30 min, a well-defined size mode had developed, with a number mode radius near 0.1-μm radius. Between 30 and 40 min, the mode structure remained stable as dilution apparently controlled the size distribution. After 40 min, essentially all of the smoke particles still had sizes below 0.5-μm radius.

The most important direct effect of smoke coagulation is to reduce the number and optical efficiency of the particles. In a series of experiments, Seader and Ou (1977) measured the "optical density" (equivalent to the specific extinction coefficient, see below) of smoke from a variety of cellulosic and polymeric materials, for flaming and nonflaming combustion, at concentrations ranging from 15 to 2750 mg/m^3, and for aging times up to several minutes or more. Smoke produced by smoldering combustion had a specific extinction coefficient of about 5 m^2/g with a maximum overall variation of a factor of 2, and smoke produced by flaming combustion, a coefficient of about 8 m^2/g with a much smaller variation. These results indicate that the optical properties of smoke are not particularly sensitive to the initial coagulation of the smoke particles. Sooty smoke, moreover, exhibits the same typical extinction and absorption coefficients for a wide range of combustion sources and aging periods (e.g., Janzen, 1980).

On the basis of this discussion of coagulation, the use of a (spherical) smoke particle mode radius of about 0.1 μm for optical calculations appears to be reasonable. Only in unusual circumstances would the average smoke particle size exceed about 0.2 to 0.3 μm due to prompt coagulation in the fire plume. In such cases, variations in γ should also be considered (and the particle morphology would play a role as well). In aged wildfire plumes, smoke particles are observed to have a mode radius of about 0.05 to 0.1 μm (the relevant data are reviewed later), which again suggests that prompt coagulation exercises only a secondary influence on average particle sizes in typical fire plumes.

Droplet Scavenging. The collection of soot and windblown charcoal debris by water droplets and ice crystals nucleated in the rising fire plume can be manifested as "black rain," which fell at Hiroshima and Nagasaki (Ishikawa and Swain, 1981). The water droplets and ice crystals that did not fall to the ground from the plume would eventually reevaporate, leaving behind their involatile cores. Many of these cores would contain one or more of the original smoke particles. In effect, the water particles can act as aggregation centers for the smoke.

Not all fire plumes form thick condensation clouds. Those that do are usually associated with large intense fires in humid environments. Because the fire itself generates immense numbers of cloud condensation nuclei, the nature of the capping cloud can be very different from a normal cumulus (L. Radke, University of Washington, private communication, 1984). Thus higher concentrations of smaller water droplets nucleated on smoke particles, ash, and windblown debris are expected. The total condensed water mass may not be significantly different from normal clouds, however (see Appendix 5-2).

Strong updrafts in the fire plume may create substantial supersaturations (Twomey, 1977). Hence smoke particles could be nucleated even though they tend to be hydrophobic when fresh (the nucleation of ~0.1-μm radius particles at supersaturations of

<1 percent usually requires that the particle have at least a small hygroscopic component (Fitzgerald, 1973)). Radke et al. (1980a) noted that cloud condensation nuclei (ccn) concentrations in the Meteotron smoke plume were comparable to those in ambient air, even though the total number of particles was enormously enhanced. These results provide evidence that fresh soot is a poor ccn.

If all of the smoke particles were to nucleate in the ascending fire plume, the concentration of droplets would exceed $10^5/cm^3$, resulting in a dense haze rather than a cloud. It is possible that only $\lesssim$ 10 percent of the submicron smoke particles are active as nuclei (Eagan et al., 1974), resulting in a more normal concentration of about 10^3 to $10^4/cm^3$ of water droplets or ice crystals (the size and number depending on the quantity of water vapor entrained into the plume).

The presence of a water cloud can alter the size (and spatial) distribution of smoke both through precipitation (by physical transport or removal of nucleated and collected smoke particles) and through coalescence of the collected smoke particles upon evaporation of the water. Precipitation removal from the fire plume is discussed under a separate heading in this chapter (see below).

The flux of aerosol particles diffusing to a cloud droplet is given by (Twomey, 1977; Pruppacher and Klett, 1978),

$$F_p = 4\pi R D_p n_p,$$

where F_p is the number of particles impinging on the droplet per second, R is the droplet radius (in centimeters), D_p is the particle diffusion coefficient (about 10^{-5} to 10^{-6} cm^2/s for particles of $\gtrsim$0.1-μm radius), and n_p is the total particle concentration (number per cubic centimeter). Typically, F_p/n_p is of the order of 10^{-7} to 10^{-8} cm^3/s for cloud droplets. If n_p is about $10^5/cm^3$, it might take about 1 h for an average droplet to collect 10 smoke particles. For a cloud droplet concentration of $1000/cm^3$, an aerosol depletion of about 10 percent would result. As the droplet evaporated, the collected smoke would combine to form a single larger smoke particle. Accordingly, cloud drop collection may on occasion be more efficient than coagulation in modifying the smoke particle size distribution in the early plume. In the long term, continuous processing of the smoke through transient water clouds would strongly affect the size distribution and trace composition of the smoke aerosols.

L. Radke (private communication, 1984) may have observed this process of smoke transformation in a cap cloud over a prescribed forest fire. He compared smoke in air that passed through the cap cloud to smoke in air that passed under the same cloud. Size distribution measurements showed that the "cloud-processed" smoke was depleted by nearly a factor of 10 of particles smaller than about 0.05-μm radius, and by a lesser factor of particles larger than about 0.5-μm radius. In the size range from 0.1 to 0.5 μm, the particle populations were similar, although the apparent number mode radius increased from about 0.05 μm to about 0.1 μm in the processed smoke. However, and most critically, the mass of submicron particles was essentially the same in both air samples (as determined from the measured volume distributions) and the optical scattering coefficients had roughly the same magnitude (within 20 percent), indicating similar optical extinction efficiencies of the smoke aerosols.

Ash and Debris Particle Scavenging. At Hiroshima, dry, dusty ash and other litter were observed to fall outside of the region of the black rain (Ishikawa and Swain, 1981). The mass of the larger debris particles in the fallout was, by implication, much greater than the mass of the attached submicron aerosol, suggesting relatively inefficient removal of smoke particles by scavenging on "fly ash." It might also be inferred from such evidence that strong fire winds are capable of lifting substantial quantities of fine dust to high altitudes in addition to the smoke, a factor that has been ignored in the present analysis.

The three potential agglomeration mechanisms for smoke particles in urban fire plumes (coagulation, cloud processing, and dust scavenging) are not explicitly taken into account in the present assessment. There is no simple way to quantify their effects on the smoke size distribution between emission at the fire and injection into the free atmosphere far downwind of the fire. However, by emphasizing observations of a variety of aged smokes, the adopted size distributions (and optical properties) already reflect the influence of these prompt physical processes.

The smoke particle size distribution continues to evolve as the plume ages. Coagulation and cloud processing act to increase the particle size, while scavenging tends to remove the largest particles preferentially (Jaenicke, 1980). Long-term coagulation has been included in some estimates of the optical and climatic effects of nuclear-generated smoke (Turco et al., 1983a,b; Crutzen et al., 1984), and in the present baseline simulations.

Optical Properties

The optical properties of smoke clouds depend on the size distribution, composition, and morphology of the smoke particles. The optical properties determine the ultimate impact of nuclear smoke emissions on solar insolation and climate. There are two basic techniques for obtaining these properties--by direct measurement and by Mie scattering computation, which is accurate for spherical particles, but only approximate for nonspherical particles represented as "volume-equivalent" spheres. In general, the assumption of volume-equivalent spherical particles tends to underestimate the absorptive power of nonspherical carbon particles (Roessler et al., 1983).

For a smoke aerosol the extinction optical depth, τ_e, at visible and thermal infrared (8 to 20 μm) wavelengths is the product of the specific extinction coefficient (or cross section), σ_e (expressed in square meters per gram of smoke), the smoke mass concentration, M (in grams per cubic meter of air), and the optical path length, L (in meters), or*

$$\tau_e = \sigma_e ML.$$

The specific extinction coefficient is simply related to the specific absorption and scattering coefficients by

$$\sigma_e = \sigma_a + \sigma_s.$$

Thus only σ_e and σ_a need to be determined. The absorption optical depth may be calculated using an equation analogous to that for the extinction optical depth.

In the visible spectrum, measurements for very pure carbon soots and carbon black yield values of $\sigma_e \simeq 10$ to 12 m^2/g and $\sigma_a \simeq 8$ to 10 m^2/g (Twitty and Weinman, 1971; Janzen, 1980; Chylek et al., 1981). At the opposite extreme, smoke from burning wood can have $\sigma_e \simeq 4$ to 6 m^2/g and $\sigma_a \simeq 0.01$ to 2 m^2/g, varying with the conditions of combustion (Patterson and McMahon, 1983). The optical coefficients of smokes generated from plastics and other polymeric compounds exhibit values intermediate between those of carbon black and wood smoke, but generally lie closer to the carbon black values (Tewarson, 1982). The extinction and absorption of smoke from hydrocarbon gas flames also approach the pure carbon values (Roessler and Faxvog, 1979). For a variety of white smokes, σ_e is observed to vary from about 2 to 5 m^2/g, and for black smokes, from about 5 to 9 m^2/g (Ensor and Pilat, 1971).

The long-wavelength infrared properties of smoke clouds may affect the heat balance of the perturbed atmosphere. In the infrared, the specific extinction coefficients of most smokes lie between 0.1 and 1 m^2/g; that is, roughly one order of magnitude below the visible extinction values (Volz, 1972; O'Sullivan and Ghosh, 1973; Roessler and Faxvog, 1979, 1980; Uthe, 1981; Vervisch et al., 1981; Bruce and Richardson, 1983). The infrared extinction is dominated by absorption.

Factors that could enhance the infrared opacity of a dispersed smoke plume include the following:

1. The presence of large windblown flyash and debris particles.
2. Water condensed as droplets.
3. Infrared active gases such as H_2O, CO_2, and unburned hydrocarbons.
4. Aggregated particles exceeding ~0.5-μm radius.
5. Highly eccentric particles.

---

*Note that the extinction of a direct beam of light obeys the law, $I/I_0 = \exp(-\tau_e)$, where I_0 and I are the incident and attenuated beam intensities, respectively. However, light scattered from the beam adds a diffuse background illumination that increases the overall transmitted intensity.

On the basis of information currently available, however, it does not appear likely that any of these factors would critically influence the long-term climatic impacts of extensive smoke clouds. For example, ash particles rapidly fall out of the clouds, and water evaporates as the cloud disperses. The persistent infrared-active gases are not produced in large enough quantities to affect the cloud infrared opacity significantly (see Appendix 5-2). Nevertheless, further analyses of the smoke infrared problem would be helpful.

For very pure carbon soot, the index of refraction inferred from optical measurements shows only a slight variation with wavelength in the visible and infrared spectrums. The real part of the index is 1.6 to 2.0, and the imaginary part* is -0.4 to -1.0 (Twitty and Weinman, 1971; Chippett and Gray, 1978; Pluchino et al., 1980; Tomaselli et al., 1981). The observed variability, particularly in the imaginary part, is probably due to varying amounts of oils (with refractive indices of ≃1.5 - 0 i) mixed with the solid carbon. Pure graphite powder can have a refractive index as large as 2.7 - 2.0 i (Tomaselli et al., 1981). For oily smoke, the imaginary index of refraction may be roughly estimated as the elemental carbon volume fraction of the smoke emission. For example, if the smoke is composed of 20 percent solid carbon by mass and 80 percent oils, then the carbon volume fraction is ≃10 percent and the smoke particle imaginary index of refraction is about -0.1.

Calculations of the optical properties of smoke require knowledge of the size distribution, refractive indices, and morphological structure of the particles. Using measured values for these physical parameters, extinction and absorption coefficients can be predicted that agree closely with the directly observed coefficients (Chylek et al., 1980; Ackerman and Toon, 1981). Such calculations also reveal the sensitivity of the optical properties to changes in the smoke parameters. For example, the specific extinction of visible light by smoke consisting of spherical particles is slightly sensitive to γ (varying by roughly a factor of 2 for limiting γ values between 1.5 and 2.5), and more sensitive to the mode radius (varying approximately as r_m^{-1} for mode radii >0.2 μm). At a wavelength of 10 μm, the overall variation in the calculated specific extinction coefficient is about a factor of 2 for all mode radii up to about 1 μm (i.e., for spherical smoke particles that lie within the Rayleigh extinction regime; Deirmendjian, 1969; Kerker, 1969). If the smoke consists of solid carbon particles suspended in droplets of oil or water, or if fine soot particles coat the surfaces of larger soil particles, the apparent specific absorption coefficient of the smoke at visible wavelengths can increase considerably (Ackerman and Toon, 1981), although its lifetime may be reduced.

---

*The index of refraction of a bulk material has a dispersive component (the "real" part) and an absorptive component (the "imaginary" part). "Real" and "imaginary" are mathematical terms that derive from the equations of light propagation. The imaginary index of refraction is generally proportional to the absorption coefficient of the material.

As was mentioned earlier, coagulation and other particle aggregation processes alter the size distribution of smoke particles in plumes and clouds. Laboratory and field measurements suggest that, for a wide range of conditions, the effect of coagulation on the optical coefficients of smoke should be less than a factor of 2 (e.g., Seader and Ou, 1977). In general, the evolution of the size distribution of the smoke particles must be taken into account in predicting long-term optical properties (Turco et al., 1983a,b). Also, most existing optical theories assume spherical particles, which probably leads to an underestimate of the optical coefficients of sooty smokes or soot/water mixtures (e.g., Pagni and Bard, 1979; Janzen, 1980).

The baseline smoke optical properties are summarized in a later section. Basically, an oily smoke is adopted with $\sigma_e = 5.5\ m^2/g$ and $\sigma_a = 2.0\ m^2/g$. Compared to observational data, these values seem to be conservative, perhaps making urban smoke look more like forest fire smoke, which is less absorbant (see below). In the baseline climate calculations of Chapter 7, the full evolution of the smoke particle size distribution and optical properties are treated using the microphysical/optical model of Turco et al. (1983a).

Forest Fire Smoke

Smoke emissions from burning forests have been measured extensively (Watson, 1951; McMahon and Ryan, 1976; Packham and Vines, 1978; Radke et al., 1978; Sandberg et al., 1979; Ward et al., 1979; McMahon, 1983; Radke et al., 1983, 1984). The particulates are typically composed of 40 to 75 percent benzene-soluble organic compounds. About 5 to 15 percent of the collected material is elemental carbon. The smoke from heading fires (moving with the wind) is yellowish to dark brown and oily, while the smoke from backing fires is black and sooty. Smoke emission factors for heading fires are ≃1 to 6 percent of the fuel burned, and for backing fires, ≃0.5 to 3 percent. The most recent observations by Radke et al. (1983, 1984) of smoke emissions from a series of carefully monitored prescribed forest burns found submicron particle emission factors averaging about 0.5 to 1.0 percent and total particle emission factors of 1 to 3 percent (but with a relatively high graphitic carbon content). It is observed that uncontrolled (wild) fires in forests produce about twice as much smoke as prescribed forest fires, per unit mass of fuel burned (Sandberg et al., 1979; McMahon, 1983). An average baseline smoke emission factor of 3 percent, consistent with the wildfire data, is assumed below.

Irrespective of the fuel type, the maximum in the smoke particle number size distribution falls near 0.05-μm radius, and in the mass distribution, near 0.15 μm (which implies a γ value of ≃1.8 to 1.9 in a log normal size distribution). Intercomparisons of size distributions measured by networks of ground samplers placed around natural fires, and by aircraft-borne instruments traversing wildfire plumes, reveal that the size distribution is generally preserved in the convective column and downwind of the fires (Sandberg et al., 1979). Even hundreds of miles from large forest fires, the smoke particles appear to remain unaltered (Watson, 1951). However, Ward et al. (1979) observed a steady increase in the abundance of optically active aerosols in the aging plume of a prescribed backing fire. While there is great variability in forest fire properties and emission rates, the greatest particle yields occur for low-intensity smoldering fires, heading fires, and fires in green and nonwoody fuels (Sandberg et al., 1979). The latest laboratory and field measurements support these general conclusions (McMahon, 1983; Patterson and McMahon, 1983; Radke et al., 1983).

Radke et al. (1983) recently detected a substantial fraction of supermicron particles in the plumes of prescribed forest fires up to several kilometers downwind. The mass mean diameter of the large particle population was about 10 to 20 μm, and the apparent emission index was occasionally as high as 0.04 g/g of fuel. Particles and embers of millimeter size were also observed. These larger particles are generally of secondary importance because of the following:

1. Low number concentrations ($<1/cm^3$).
2. Short atmospheric residence times (against sedimentation and washout).
3. Negligible optical effects and small infrared effects.

While of less interest here, the smoke emissions from prescribed fires in grass stubble and straw amount to about 1 percent of the material burned (Boubel et al., 1969). Wild grass fires could emit twice this amount.

The observed specific scattering coefficient of forest fire smoke at visible wavelengths is 2.6 to 7.0 m^2/g (Evans et al., 1977; Tangren, 1982; Radke et al., 1983). The specific absorption coefficient of smoke generated by prescribed burns of palmetto leaves and pine needles is found to lie in the range of 0.04 to 2.8 m^2/g (Patterson and McMahon, 1983). These low optical absorptivities are consistent with a graphitic carbon mass fraction of about 5 to 10 percent in the smoke. Radke et al. (1983) measured graphitic carbon fractions of 10 to 15 percent in smoke from prescribed forest fires. Some wildfires might have considerably higher solid carbon contents and optical absorptivities (Sandberg et al., 1979; McMahon, 1983).

While the infrared properties of wildland smokes are not well established, it has been observed that solar near-infrared radiation is much less attenuated by forest fire smoke than is visible radiation (Wexler, 1950). Such a result is consistent with smoke plumes dominated by submicron particles. The presence of fly ash (Radke et al., 1983) and water vapor (Appendix 5-2) in the plumes would increase their infrared opacity, particularly near the fire sources. However, over the long term, the infrared effects of wildfire smoke would appear to have little significance.

The optical anomalies produced by forest fire smoke plumes provide some information on the properties of the smoke particles. By far, the most commonly reported optical effects are red and yellow-green suns (e.g., Lyman, 1918), yellow skies, dry fogs, and dark days (Plummer, 1912). These effects are all consistent with heavy emissions of submicron, oily smoke particles like those that have been collected in wildfires.

Occasionally, blue suns and moons are observed through smoke clouds. The most notable case is the Alberta, Canada, forest fire of September 1950 (Wexler, 1950). The blue sun phenomenon has been interpreted in terms of an aerosol composed of particles of about 0.5-μm radius (Watson, 1952). The aerosol could be formed by water vapor condensed on smoke particles. Even though Watson (1951) measured smoke particle sizes of about 0.05 μm in air that had descended to the ground from the Alberta smoke plume, the water that may have been condensed aloft could have evaporated as the air subsided and warmed. An alternative explanation for the appearance of blue suns involves the contrast of an occluded solar disk against a smoky sky and the color sensitivity of the eye. However, this argument may not apply to the Alberta smoke pall (Wexler, 1950). In any case, blue suns viewed through fire plumes are rare.

Satellite imagery of wildfire smoke plumes at visible and infrared wavelengths is available (Matson et al., 1984), and might be analyzed to determine the optical properties of the smoke clouds as they disperse over the course of several days.

Fire Burning Times

It is assumed that urban fires would burn out in about 1 day, with the period of most intense burning confined to several hours (FEMA, 1982). Fires in dense fuel arrays might persist for several days. Hence most of the smoke injection following a nuclear exchange would occur over a relatively short time (assuming that the exchange itself would be executed within a few days). Except in cases of unusually dry and windy weather, nuclear forest fires would probably burn out within 1 week.

If the nuclear detonations were distributed over a period of weeks to months, the principal findings discussed here need to be revised. Such a concept of controlled nuclear war fighting is not widely accepted by nuclear strategists (see Chapter 3).

Smoke Injection Altitudes

The heights at which nuclear smoke clouds would stabilize can be estimated by using observational data and buoyant plume theory. Anecdotal information concerning the ascent of large fire plumes is reviewed in Appendix 5-1. During intense wildfires and prescribed forest fires, plumes of smoke generally reach altitudes exceeding 1 km and can easily rise to 6 km or more (Wexler, 1950; Taylor et al., 1973; Eagan et al., 1974; Packham and Vines, 1978; Radke et al., 1978, 1983). In the late, smoldering stages of wildfires, the smoke is generally confined to the lowest kilometer of the atmosphere. However, the smoke emission during this phase of the fire is of secondary interest here.

Large-scale (>1-km diameter) urban nuclear fires probably would deposit much of their smoke in the "free" troposphere above the planetary boundary layer (Miller and Kerr, 1965). Deposition at these heights is supported by World War II experiences (e.g., Hamburg and Dresden) and by simulations of urban fire plume dynamics (e.g., Brode et al., 1982; Larson and Small, 1982a,b; Carrier et al., 1984; Cotton, 1984).

Smoke that is injected at high altitudes tends to remain aloft as it disperses over large areas. (Thus the invention and extensive use of smokestacks to reduce surface pollution.) When the smoke can penetrate the boundary layer, or a temperature inversion at any level, it may remain aloft for very long periods. An excellent example of this is afforded by the Alberta, Canada, forest fire of September 1950 (Smith, 1950; Wexler, 1950). Most of the Alberta smoke remained well above an altitude of several kilometers, and eventually reached the tropopause, as it dispersed over an area of about 10^7 km^2 during the course of a week. In the past, there have been many fires that created dark days and dry fogs over distances of hundreds to thousands of kilometers (Plummer, 1912).

The theory of buoyant plumes in a stratified atmosphere is described by Briggs (1975). He reviewed observational data and theoretical treatments of the problem for heat sources up to 1000 MW. For smoke rising in still air, the height of the center of the plume may be estimated as

$$z_c \simeq 1/5Q^{1/4},$$

where z_c is in kilometers, and Q is the heat source in megawatts. This relation holds for a constant temperature lapse rate of 6.5°C/km (an average value) and horizontal wind speeds of <5 m/s. In stronger wind fields, the plume center height is given approximately by

$$z_c \simeq 1/9Q^{1/3}U^{-1/3},$$

where U is the wind speed in meters per second. The vertical thickness of the plume is typically <50 percent of the center height. Measured plume heights and energy release rates (for example, by Taylor et al. (1973) in an intense forest-slash fire) may be correlated very well with this simple plume theory.

If the plume equations are extrapolated to large-scale urban fires, a rough estimate of potential smoke injection heights can be obtained (Manins, 1984). Q is determined in this case by the relation,

$$Q \simeq Am_0f_be_c/t_b \times 10^4,$$

where A is the fire area in square kilometers, m_0f_b is the mass of combustible material burned per unit area (in grams per square centimeter), e_c is the heat of combustion of the fuels (20,000 J/g), and t_b is the burning time. For a city fire with A = 100 km^2, m_0f_b = 3 g/cm^2, and $t_b = 10^4$ s, Q = 6 x 10^6 MW, and $z_c \simeq 10$ km.

Carrier et al. (1984) modeled the hydrodynamics of a large-scale fire convective column and concluded that a 1×10^6 MW fire could generate a plume reaching 8 to 10 km in height. They suggested that, under special circumstances, the plume could go even higher. The urban fires expected in a full-scale nuclear war would be unprecedented in number and size. Based on observations and simulations of buoyant plumes, it appears that smoke from nuclear fires could be injected well into the free troposphere, and even up to the tropopause (Cotton, 1984; Manins, 1984).

There are a number of factors that could decrease the fire plume heights, and others that could increase them. Decreases could be caused by the following:

1. Stiff crosswinds and turbulence.
2. Strong temperature inversion and atmospheric stability.
3. Infrared radiative cooling under nighttime conditions.
4. Low-level atmospheric divergence.
5. Slow burning.

Factors that could increase plume heights include the following:

1. High ambient surface humidity (and latent heat release).
2. Marginal atmospheric stability or conditional instability.
3. Solar absorption and heating in the plume.
4. Low-level atmospheric convergence.
5. Rapid burning.

Davies (1959) describes the plume of an oil refinery fire that apparently acquired considerable buoyancy by solar absorption. The additional energy seems to have accelerated the dispersion of the plume.

The height distribution of the injected smoke, summed over all urban and forest fires, depends on the statistical distribution of the plume heights, each of which depends on fire area and intensity and meteorological conditions, among other things. There is no reason to believe, a priori, that the smoke would be preferentially injected near the ground or at high altitudes. However, common experience even with rather small fires suggests that the smoke would be channeled upward by the fire convection column to a distinct stabilization height. Because forest fires generate only a small fraction (about 17 percent) of the total smoke emission in the present baseline case, attention is focused here on urban fires.

To make a crude estimate of the height profile of smoke injection, a reasonable (and convenient) statistical distribution of urban fire areas may be chosen. We assume that the <u>total</u> area of fires which fall into a size range $A \rightarrow A + dA$ is proportional to $dA/A^{3/4}$, where A is the area of an individual fire or fire complex. This distribution strongly emphasizes the number of small fires, even though the overall fire area is dominated by the larger fires. Such a distribution may be conservative in a world having only about 1000 major urban centers. The vertical thickness of the stabilized smoke plume can be taken as proportional to the plume center height, z_c (Briggs, 1975). If the

particles could grow uniformly by water vapor condensation only to an average size of about 1-μm radius. The coalescence of water droplets is extremely sensitive to droplet size, and 1-μm droplets normally require many hours to form precipitation (Twomey, 1977).

In fires that are intense enough to generate strong convection, the fire winds also sweep up large quantities of supermicron debris particles (soil, ash, and char). Windblown debris has been observed in the plumes of forest fires (Radke et al., 1983) and was seen to fall out of the clouds at Hiroshima, Nagasaki, and other large World War II fires. This debris could preferentially nucleate to form rain and cloud drops. It is possible, for example, that the most significant contribution to the recorded "black rain" events of World War II was the prompt washout and rainout of charred, windblown fire debris. Water vapor nucleation and subsequent condensational growth might directly affect up to about 10 percent of the submicron smoke particles in strong plume updrafts with rapid cooling rates; as already noted, freshly formed smoke particles tend to be hydrophobic, or water repellant, and thus are inherently poor cloud nuclei (Charlson and Ogren, 1982).

Smoke can also be collected by cloud droplets that nucleate, grow, and later coalesce into raindrops. However, as brought out in the previous discussion of the smoke size distribution, the lifetime of the submicron smoke particles against diffusional collection (in a cloud of about 1000 10-μm droplets per cubic centimeter) is likely to be an hour or more. On the other hand, the black rain is probably formed in the rising convective column within minutes of the smoke emission. This is demonstrated by the fact that the mass centroid of the induced rainfall at both Hiroshima and Nagasaki was over, but just downwind of, the fire area (Ishikawa and Swain, 1981).

Smoke particles can be scavenged by rapidly growing ice crystals and water droplets through phoretic, inertial, and electrostatic forces (Pruppacher and Klett, 1978). However, the absolute efficiencies for the collection of submicron particles is not well defined. In one particularly interesting study, Prodi (1983) noted that, while submicron particles of sodium chloride, which is quite hygroscopic, were readily scavenged by rapidly growing ice crystals, submicron droplets of Caranuba wax, which is hydrophobic, were essentially unaffected.

In the case of inertial capture of submicron aerosols by precipitation (rain and ice), the collection efficiencies are exceedingly low (Pruppacher and Klett, 1978), because the small particles tend to be deflected by the airstream that flows around a falling hydrometeor, thus preventing direct collisions.

Evaporating cloud and precipitation drops and ice crystals would collect submicron particles principally by thermophoresis, perhaps aided by electrostatic attraction, (Pruppacher and Klett, 1978). This process would apply, for example, to raindrops falling through a subsaturated air layer, or to cloud water or ice in an evaporating cumulus cap cloud or anvil. However, since the water is evaporating, transformation rather than removal of the smoke is generally implied.

Field measurements of the collection efficiencies of aerosols by rain and cloud drops are often 10 to 100 times larger than theory indicates (e.g., Radke et al., 1980b). Laboratory measurements of collection efficiencies are also generally larger than theoretical values (e.g., Leong et al., 1982). Electrical forces may play a significant role in increasing the collection efficiency (Wang et al., 1978), particularly in the case of violent updrafts where electrification conditions similar to those in thunderstorms might develop. Observations of large droplet collection efficiencies have been discussed recently by Slinn (1983), but their cause remains in dispute.

The cloud/precipitation collection efficiencies are lowest for aerosols in the size range from about 0.1 to 1.0 μm, the so-called "Greenfield gap" (Greenfield, 1957). Barlow and Latham (1983) estimated a submicron aerosol half-life in a 2 mm/h rainfall (at relatively humidities of 50 to 70 percent) of 7 to 70 h, corresponding to collection efficiencies* of 10^{-2} to 10^{-3} (for charged and uncharged particles, respectively). In a thunderstorm, with an estimated raindrop collection efficiency of 0.1, Barlow and Latham estimated an aerosol half-life of about one-half hour.

Turbulence could enhance the rate of smoke scavenging by cloud drops. However, in the region of the fire plumes where condensation and precipitation occur, the turbulence field should not be any more intense than in natural convective clouds. Rosenkilde and Serduke (1983) showed that such turbulence would not significantly augment the aerosol removal rate.

Even after the smoke particles are captured in cloud droplets, they must still be removed in precipitation. There are only a few well-documented cases of significant prompt rainfall from fire plumes (e.g., Hiroshima). Mordy (1960) describes the singular lack of induced precipitation during the massive burning of sugar cane fields in Hawaii. In an air parcel rising over a large fire, the time available for the formation of precipitation is quite short (less than one-half hour). The cloud droplets may also be unnaturally small in size. Accordingly, the situation in a fire plume is fundamentally different from that in a natural convective/precipitation system, in that the buoyant instability is created in large part by "dry" heat, and the rising air is seeded with unusually high numbers of cloud condensation nuclei. Typically, the ambient atmosphere in the vicinity of a nuclear fire would be found in a relatively stable initial state (i.e., without strong local convection and storms). During natural precipitation events, the half-lifetime of cloud droplets against removal as rain lies in the range of 10^3 to 10^4 s (Pruppacher and Klett, 1978). In fire plumes, longer droplet lifetimes could be expected.

For the baseline calculations, it is assumed that, on the average, 50 percent of the smoke emissions from urban fires is promptly removed

---

*The collection efficiency is defined as the fraction of aerosols within the volume traced out by a falling raindrop that collide with and adhere to the drop.

by precipitation in the fire plumes. This estimate is conservative with respect to the previous discussion, which suggests that precipitation would not occur at all fires, and that removal efficiencies for submicron smoke particles could be quite low. Prompt scavenging reduces the overall baseline smoke emission factor from 0.04 g/g to 0.02 g/g for urban fires. The latter figure is utilized below to make smoke emission estimates. In reality, smoke scavenging might be concentrated in the most intense fires with the tallest convective columns, under conditions of high ambient surface humidity. However, lacking detailed quantitative information on the simultaneous probability of all of these conditions, the precipitation removal is applied uniformly to all urban smoke plumes. Some smoke would also be redistributed by the cloud and precipitation processes, but this effect is ignored.

Finally, it should be mentioned that the long-term removal of dispersed smoke by precipitation is explicitly taken into account in the baseline climate calculations (Chapter 7). Global removal by wet deposition is the principal sink for the smoke that escapes from the fire plumes (Jaenicke, 1980; Charlson and Ogren, 1982; Turco et al., 1983c). Removal processes and typical atmospheric lifetimes for smoke particles are discussed in Chapter 7.

ESTIMATING SMOKE EMISSIONS IN A MAJOR NUCLEAR EXCHANGE

Baseline Estimates

The nuclear war scenarios considered in this report are highly generalized. No detailed information is given regarding explosion yields or heights of burst for specific targets, or the duration of the exchange. The baseline scenario utilizes 6500 Mt, of which 1500 Mt is targeted on urban areas (Chapter 3), where key military and industrial targets are located (e.g., Kemp, 1974; Ball, 1981). Moreover, it is reasonable to assume that a substantial fraction of the remaining 5000 Mt would be detonated over targets near forests, brush, and grass lands. Although the baseline case in this study does not postulate a season for the exchange, for the purposes of calculation of wildfire smoke, summer season values are used (urban fires, the principal source of smoke, would be largely independent of seasonal variation).

The baseline smoke emission estimates are given here, and excursions from the baseline are discussed in the next subsection. As outlined earlier, it is assumed that 250,000 km^2 of urbanized area is partially burned, which corresponds to 50 percent of the total urbanized area in the countries at war. Such an area could be ignited by 1500 Mt of air bursts, assuming an average ignition area of ≈ 250 km^2/Mt, no fire spread, one-third overlap of ignition zones, and no spreading beyond the 20 cal/cm^2 ignition zone. Within the fire area, the average burden of combustible material is taken to be 4 g/cm^2, and three-quarters of this is assumed to burn, in accordance with other estimates of urban fire damage in a nuclear attack (e.g., Miller et al., 1970; FEMA, 1982; Brode and Small, 1983). Finally, the net smoke emission factor is assumed to be 0.02 g/g (grams of smoke per gram of fuel consumed) after scavenging and removal by coagulation and condensation processes in the convective fire plumes are taken into account (50 percent removed). Multiplying the appropriate factors together, the total urban smoke emission amounts to ≈150 Tg (1.5×10^{14} g).

Forest fires are also estimated to burn 250,000 km^2 (i.e., roughly the area of irradiation at ≥20 cal/cm^2 by 1000 Mt of air bursts). The basis for this estimate is discussed earlier in this chapter. The fuel consumed in forest fires is taken to be 0.4 g/cm^2 (about 20 percent of the typical fuel loading), and the net smoke emission factor is taken to be 0.03 g/g, both values based on observations. Brush and grass fires, whose emissions are smaller per unit area burned, are not explicitly included in the analysis. The total forest fire smoke emission is then ≈30 Tg. In winter, wildfire emissions might be reduced to a few teragrams; however, because urban fires contribute much more soot, the total emission would be reduced by no more than 20 percent.

The composition and optical properties of the smoke in the baseline model must also be specified. Even though urban fires dominate the aggregate smoke emission in the baseline case, with potential soot fractions of up to 90 percent, it is assumed that the average graphitic carbon fraction is only 20 percent (compared to ~10 percent in forest fire smoke). The smoke particle number size distribution is taken to be log normal with a number mode radius* of 0.1 μm and γ = 2.0; the effective particle density is 1 g/cm^3. The smoke index of refraction is 1.55 - 0.1 i. At visible wavelengths the corresponding smoke specific extinction coefficient can be taken as 5.5 m^2/g, and the specific absorption coefficient as 2.0 m^2/g. The smoke infrared extinction and absorption coefficients (at 10 μm) are both roughly 0.5 m^2/g. These physical constants provide a consistent set for optical (Mie) calculations.

Because the selected baseline optical extinction and absorption coefficients are much smaller than typical values for sooty (urban) smokes, the effect of "aging," which can reduce the optical efficiency of the smoke, may be neglected in carrying out approximate optical-effects simulations. The optical efficiency per unit mass of smoke is otherwise expected to decline with time.

The smoke parameters for the baseline nuclear war scenario are summarized in Table 5.7. The total estimated smoke emission is 180 Tg, caused by roughly 30 percent of the nuclear explosions. The estimated smoke emissions are very uncertain, however; some of the sources of uncertainty are discussed below.

The total quantity of combustibles consumed in the baseline war scenario is 8500 Tg (7500 Tg in urban fires and 1000 Tg in forest fires). For the urban flammables, about 5000 Tg of cellulosics, 1500 Tg of liquid fossil organics, and 1000 Tg of industrial organochemicals, plastics, polymers, rubbers, resins, etc., are burned. The

---

*For volume-equivalent spherical particles.

TABLE 5.7 Fire and Smoke Parameters in the Present Nuclear War Analysis

	Baseline	Excursions[a]
Urban fire smoke emission, Tg	150	20-450
Forest fire smoke emission, Tg	30	0-200
Total smoke emission, Tg	180	20-650
Tropospheric injection, Tg/km	20 (0-9 km)	1.5-53 (0-12 km)
Stratospheric injection, Tg/km	0	1 (12-20 km)
Urban fire area, km^2	250,000	125,-375,000
Urban fuel consumption, g/cm^2	3.0	1.5-3.0
Urban smoke emission factor,[b] g/g	0.02	0.01-0.04
Urban fire duration, days	$\lesssim 1$	--
Forest fire area, km^2	250,000	0-1,000,000
Forest fuel consumption, g/cm^2	0.4	0.4
Forest smoke emission factor, g/g	0.03	0.02-0.05
Forest fire duration, weeks	$\lesssim 1$	--
Smoke composition (by mass)	20% graphitic carbon, 80% oils	5-50% graphitic carbon
Smoke refractive index (visible)[c]	1.55-0.10 i	1.5-0.02 i to 1.7-0.30 i
Smoke particle number median size, μm	0.10	0.05-0.5
Smoke particle log normal width, γ	2.0	2.0
Smoke specific extinction (visible),[c] m^2/g	5.5	2.0-9.0
Smoke specific absorption (visible),[c] m^2/g	2.0	1.0-6.0
Smoke specific absorption (infrared), m^2/g	0.5	0.2-5.0

[a]Some values are given only to illustrate the range that is plausible, and are not discussed specifically in the text.
[b]Average value after 50 percent prompt scavenging in the convective fire columns.
[c]At a nominal wavelength of 550 nm.

corresponding total energy release is about 5 x 10^{19} cal, or 50,000 Mt, assuming an average heat of combustion of 6000 cal/g. (Note, by comparison, that one day's solar insolation amounts to about 3,000,000 Mt of energy.) The energy release drives the buoyancy of the fire plumes and may create strong surface winds. Because the initial nuclear detonations over cities would pulverize large quantities of masonry and plaster into fine dust, it is likely that a significant burden of submicron particulates would be drawn up into the fire plumes. Even if only 1000 tons of fine (submicron) dust were raised for each megaton of thermal energy released, the dust injection could total 30 Tg. However, because there are few data pertaining to this source of particulates, it is ignored in the baseline assessment; future consideration seems worthwhile.

As was discussed earlier, the smoke mass insertion is assumed to be uniform with height between the ground and 9-km altitude, and to occur over a period of several days to 1 week.

Excursions from the Baseline Case

In order to place some limits on the possible range of smoke emissions in the baseline scenario, reasonable excursions of the fire parameters are investigated. These excursions are not meant to represent an absolute range of possibilities, but a range that seems to be consistent with current scientific knowledge. In the case of urban fires, the area burned is varied between 25 percent and 75 percent of the urbanized area of the NATO and Warsaw Pact countries (neglecting possible urban damage in other industrialized nations such as China and Japan), the net smoke emission factor is varied between 0.01 g/g and 0.04 g/g, and the fuel burden is varied between 2 g/cm^2 and 4 g/cm^2. None of these assumptions appears to be extreme. The resulting urban smoke emission varies from ≈20 Tg to ≈450 Tg. This range of emissions is in rough accord with the range estimated by Broyles (1984). In the case of forest fires, it is assumed, on the low side, that no smoke emissions would occur. On the high side, a fourfold increase in the burned area and a smoke emission factor of 0.05 g/g are assumed, yielding a forest smoke emission of ≈200 Tg. Accordingly, the present estimate of a potential range of smoke emissions following the baseline nuclear exchange is ≈20 to ≈650 Tg. This is not an uncertainty range for the emission, but an excursion range based on plausible parameter variations. Sources of uncertainty in these estimates are discussed in the next section.

Because it is possible that the smoke plumes of massive urban fires would penetrate into the stratosphere, it is worthwhile to consider the implications of smoke injections in the lower stratosphere. The injection of up to 10 Tg of smoke (just over 5 percent of the baseline estimate) between 12 and 20 km at northern mid-latitudes may be assumed. Even though such an injection is not included in the baseline calculation, it represents a potentially interesting excursion (Turco et al., 1983a,b).

Turco et al. (1983a,b) pointed out that massive smoke emissions would be possible in nuclear exchanges that involved only a limited total yield detonated over or near major urban centers. This conclusion is based on the observation that most urban areas tend to have dense "cores" in which combustible materials are concentrated. Thus about 100 Mt (say, in 50- and 100-kt weapons) would be sufficient to attack all of the major urban centers in the NATO and Warsaw Pact countries. Such a purposefully destructive strategy is currently thought to be unlikely. However, an equivalent result is possible. For a scenario of any size in which 100 Mt of explosions were to burn an urban area of 25,000 km^2 (about 50 percent of the city cores of the combatant nations), consume 20 g/cm^2 of combustibles, and emit 2 percent (net) of the burned mass as particulate in the process, $\approx$100 Tg of smoke would be generated. This is similar to the baseline urban smoke emission of 150 Tg. However, the emission would be patchier for a longer time in the 100-Mt case due to a reduced number of smoke sources.

In accordance with the estimates presented above, one may deduce that smoke emissions from nuclear-initiated wildfires scale very roughly with the total yield of the exchange, including tactical weapons, and are very sensitive to season, with maximum emissions in summer and early fall and minimum emissions in winter. Smoke production by urban fires, on the other hand, may be rather insensitive to total yield, if the urban centers, or the military and industrial sites within urban zones, are systematically targeted. The effect of seasonal and meteorological conditions on nuclear urban fires (as with everyday urban fires) is also less important, owing to the general protection of urban combustibles from the weather.

Optical Depth Excursions

Given the range of smoke emissions just described and the possible variations in smoke optical properties summarized in Table 5.7, ranges of <u>average</u> optical depths (at visible wavelengths) can be calculated. Assuming hemispherical dispersion of the smoke, the baseline extinction optical depth is 4, and the absorption optical depth is 1.4. If the baseline smoke optical constants are accepted, the hemispherical extinction optical depth can range from 0.44 to 14.3; the optical depth for smoke confined to the northern mid-latitudes can range from 1 to 36. The corresponding absorption optical depths are about one-third of these values. Taking into account possible variations in the smoke optical constants, the optical depth range is even greater.

Significant radiative and climatic perturbations might be expected whenever the hemispheric-scale optical depth is $\gtrsim$1. Volcanoes, which generate only nonabsorbing aerosols, can produce noticeable global disturbances at optical depths of about 1 (Chapter 8). Accordingly, the major segment of the optical depth range derived above can lead to serious environmental effects (see Chapter 7 for an exposition of these effects). However, given the large plausible ranges of fire and smoke parameters, subcritical optical depths clearly lie within the range of uncertainty.

UNCERTAINTIES

Uncertainties are recognized in each of the key parameters pertaining to fires and smoke emissions in a nuclear war. Although only very rough estimates of the uncertainties may be deduced, even these may be useful in evaluating the weaknesses in current knowledge. Accordingly, a subjective assessment of uncertainties, based on consideration of the limited set of data available to the committee, is spelled out below.

1. The areal extent of nuclear urban fires per megaton of yield (factor of 2 to 3). Potential overlap of fire zones, and fire spread, dominates the uncertainty.
2. Quantities and distributions of flammable materials in cities and surrounding areas (factor of 3 in the average central-city fuel burden, factor of 2 in the average suburban fuel burden, factor of 3 in the worldwide urban-area average fuel burden).
3. Urban smoke emissions per unit mass of combustible loading (factor of 2 in the fraction of fuel burned in urban nuclear fires, factor of 2 to 3 in the quantity, or mass, of smoke generated per unit mass of material burned, factor of 3 in the graphitic carbon mass fraction, factor of 2 to 3 in the mean particle size, and factor of 1.5 in the average particle bulk density).
4. Optical (visible wavelength) properties of urban fire smoke (factor of 2 in the specific extinction and scattering coefficients (square meters per gram), factor of 3 in the specific absorption coefficient (square meters per gram), factor of 3 in the imaginary part of the refractive index).
5. Infrared properties of urban fire smoke (factor of 3 in the late-time specific extinction/absorption coefficient; factor of 5 in the early-time extinction/absorption coefficient which may be controlled by condensed water and fly ash).
6. The areal extent of nuclear forest fires (factor of 3 to 4, neglecting sensitivity to the explosion scenario).
7. Forest fire smoke emissions per unit area burned (factor of 2 to 3 in the fraction of biomass fuel consumed, factor of 2 in the mass of smoke emitted per unit mass of fuel burned, factor of 3 in the graphitic carbon mass fraction, factor of 2 to 3 in the mean particle size, and factor of 1.5 in the average particle bulk density).
8. Optical (visible wavelength) properties of forest fire smoke (factor of 1.5 to 2 in the specific extinction and scattering coefficients (square meters per gram), factor of 3 in the specific absorption coefficient (square meters per gram), factor of 3 in the imaginary part of the refractive index).
9. Infrared properties of forest fire smoke (factor of 2 to 3 in the specific extinction/absorption coefficient at intermediate and late times).
10. Heights of smoke plumes from mass nuclear urban and forest fires (factor of 1.5 to 2 in both cases).
11. Extent of precipitation scavenging (black rain) and coagulation in the most intense fire plumes (the overall precipitation scavenging efficiency could vary from 25 to 75 percent; the reduction

of the optical extinction and absorption coefficients by prompt coagulation in the densest plumes could vary from 20 to 50 percent).

12. Quantity of submicron masonry dust raised in urban fire plumes following pulverization of buildings by nuclear blast (injection of 0 to 10^5 tons/Mt of explosive yield); the extent of smoke production from burning aluminum and other "nonflammable" materials in very intense fires is unknown.

13. Effect of massive smoke emissions on the subsequent meteorology and particle removal rates (factor of 3 to 10; see Chapter 7).

The uncertainty factors defined above cannot simply be multiplied to estimate absolute ranges of equally likely values for composite parameters such as smoke emissions and optical depths. The factors do not correspond to intervals of statistical significance, in which the central (or baseline) values are the most probable values. Because the various smoke parameters are largely uncorrelated, the uncertainty in combinations of the parameters must be deduced by statistical means. A precise determination of the overall uncertainty in the smoke emission and optical depth estimates cannot be made at this time, because the nature of the statistical dispersion has not yet been ascertained.

The propagation of uncertainty into the radiative transfer and climate calculations has an exponential component, because those calculations involve terms of the form, $e^{-\tau}$. Using the present baseline case as a reference, an increase in the smoke emissions would have less impact than a decrease, inasmuch as the light absorption by the smoke is already about 90 percent, averaged over the northern hemisphere. The duration of significant effects would be prolonged, however. Patchiness, or light leakage through "holes" in the smoke clouds, also has an exponential dependence. Nevertheless, average smoke optical depths of even ~1 would still imply major perturbations of the postwar environment (for example, volcanic scattering optical depths ~1 can produce significant climate anomalies). The climatic aspects of the light transmission problem are discussed in Chapter 7.

Turco et al. (1983a,b) carried out a large number of sensitivity tests in which the physical parameters of smoke and dust and the explosion scenarios were varied to investigate the nature of the uncertainty in the smoke emission, light transmission, and climate variation. They concluded that as many uncertain factors could act to aggravate the effects as could act to ameliorate them.

SUMMARY

A full-scale nuclear exchange of 6500 Mt, involving a variety of military and urban targets, would ignite numerous fires and could generate as much as 180 Tg of smoke. Considering the substantial uncertainties involved in estimating the smoke emission, however, the plausible range of emissions extends from 20 to 650 Tg. The optical properties of the dispersed smoke clouds have been deduced principally from observational data. At visible wavelengths, a specific extinction coefficient of 5.5 m^2/g and a specific absorption coefficient of 2.0 m^2/g are selected for optical and climate calculations. The infrared extinction coefficient is an order of magnitude smaller than the visible extinction coefficient. The baseline optical absorptivity is conservative (on the low side), in view of the strong absorption of light by typical sooty smokes. Even so, the implied disturbances in solar transmission on a global scale appear to be serious.

REFERENCES

Ackerman, T.P., and O.B. Toon (1981) Absorption of visible radiation in atmosphere containing mixtures of absorbing and nonabsorbing particles. Appl. Opt. 20:3661-3668.

Ayers, R.U. (1965) Environmental Effects of Nuclear Weapons. Vols. 1-3. Report HI-518-RR. Harmon-on-Hudson, N.Y.: Hudson Institute.

Backovsky, J., S.B. Martin, and R. McKee (1982) Experimental Extinguishment of Fire by Blast. Report PYU-334. Menlo Park, Calif.: Stanford Research Institute. 97 pp.

Baldwin, R. (1968) Flame merging in multiple fires. Combust. Flame 12:318-324.

Ball, D. (1981) Can Nuclear War Be Controlled? Adelphi Paper 169. London: International Institute Strategic Studies.

Bankston, C.P., B.T. Zinn, R.F. Browner, and E.A. Powell (1981) Aspects of the mechanisms of smoke generation by burning materials. Combust. Flame 41:273-292.

Barlow, A.K., and J. Latham (1983) A laboratory study of the scavenging of sub-micron aerosol by charged-raindrops. Pages 551-560 _in_ Precipitation Scavenging, Dry Deposition and Resuspension, edited by H.R. Pruppacher, R.G. Semonin, and W.G.N. Slinn. New York: Elsevier.

Baum, H.R., and G.W. Mulholland (1979) Coagulation of smoke aerosol in a buoyant plume. J. Colloid. Interface Sci. 72:1-12.

Benech, B. (1976) Experimental study of an artificial convective plume initiated from the ground. J. Appl. Meteorol. 15:127-137.

Benech, B., J. Dessens, C. Charpentier, H. Sauvageot, A. Druilhet, M. Ribon, P.V. Dinh, and P. Mery (1980) Thermodynamic and microphysical impact of a 1000 MW heat-released source into the atmospheric environment. Pages 111-118 _in_ Proceedings of the Third WMO Scientific Conference on Weather Modification. Geneva: World Meteorological Organization.

Boubel, R.W., E.F. Darley, and E.A. Schuck (1969) Emissions from burning grass stubble and straw. J. Air Pollut. Control Assoc. 19:497-500.

Briggs, G.A. (1975) Plume rise predictions. Pages 59-111 _in_ Chapter 3 of Lectures on Air Pollution and Environmental Impact Analyses. Boston, Mass.: American Meteorological Society.

Brode, H.L., and R.D. Small (1983) Fire Damage and Strategic Targeting. PSR Note 567. Santa Monica, Calif.: Pacific Sierra Research Corp. 58 pp.

Brode, H.L., D.A. Larson, and R.D. Small (1982) Time-Dependent Model of Flows Generated by Large Area Fires. PSR Note 483. Santa Monica, Calif.: Pacific Sierra Research Corp.

Broido, A. (1960) Mass fires following nuclear attack. Bull. At. Sci. 16:409-413.

Broyles, A.A. (1984) Smoke generation in a nuclear war. Submitted to Am. J. Phys.

Bruce, C.W., and N.M. Richardson (1983) Propagation at 10 μm through smoke produced by atmospheric combustion of diesel fuel. Appl. Opt. 22:1051-1055.

Butcher, S.S., and M.J. Ellenbecker (1982) Particle emission factors for small wood and coal stoves. J. Air Pollut. Control Assoc. 32:380-384.

Cadle, R.D. (1972) Composition of the stratospheric sulfate layer. Eos 53:812-820.

Calcote, H.F. (1981) Mechanisms of soot nucleation in flames--A critical review. Combust. Flame 42:215-242.

Carrier, G.F., F.E. Fendell, and P.S. Feldman (1982) Firestorms. TRW Report 38163-6001-UT-00. Redondo Beach, Calif.: TRW Systems. 83 pp.

Carrier, G.F., F.E. Fendell, and P.S. Feldman (1984) Big fires. To appear in Combust. Sci. Technol.

Chandler, C.C., T.G. Storey, and C.D. Tangren (1963) Prediction of Fire Spread Following Nuclear Explosions. Research Paper PSW-5. Washington, D.C.: U.S. Forest Service. 110 pp.

Charlson, R.G., and J.A. Ogren (1982) The atmospheric cycle of elemental carbon. Pages 3-18 _in_ Particulate Carbon: Atmospheric Life Cycle, edited by G.T. Wolff and R.L. Klimisch. New York: Plenum Press.

Chippett, S., and W.A. Gray (1978) The size and optical properties of soot particles. Combust. Flame 31:149-159.

Church, C.R., J.T. Snow, and J. Dessens (1980) Intense atmospheric vortices associated with a 1000 MW fire. Bull. Am. Meteorol. Soc. 61:682-694.

Chylek, P., V. Ramaswamy, and M.K.W. Ko (1980) Effects of aerosol size distributions on the extinction and absorption of solar radiation. Pages 225-235 _in_ Environmental and Climatic Impact of Coal Utilization, edited by J.J. Singh and A. Deepak. New York: Academic Press.

Chylek, P., V. Ramaswamy, R. Cheng, and R.G. Pinnick (1981) Optical properties and mass concentration of carbonaceous smokes. Appl. Opt. 20:2980-2985.

Cotton, W.R. (1984) A simulation of cumulonimbus response to a large firestorm--Implication to a nuclear winter. Submitted to Science.

Crutzen, P.J., and J.W. Birks (1982) The atmosphere after a nuclear war: Twilight at noon. Ambio 11:114-125.

Crutzen, P.J., C. Brühl, and I.E. Galbally (1984) Atmospheric effects from post-nuclear fires. Climatic Change, in press.

Culver, C.G. (1976) Survey Results for Fire Loads and Live Loads in Office Buildings. Report NBS BSS-85. Gaithersburg, Md.: National Bureau of Standards. 157 pp.

Dasch, J.M. (1982) Particulate and gaseous emissions from wood-burning fireplaces. Environ. Sci. Technol. 16:639-645.

Davies, R.W. (1959) Large-scale diffusion from an oil fire. Pages 413-415 in Atmospheric Diffusion and Air Pollution, edited by F.N. Frenkiel and P.A. Sheppard. New York: Academic Press.

Day, T., D. MacKay, S. Nadeau, and R. Thurier (1979) Emissions from in situ burning of crude oil in the Arctic. Water Air Soil Pollut. 11:139-152.

Deirmendjian, D. (1969) Electromagnetic Scattering on Sperical Polydispersions. New York: Elsevier. 290 pp.

Dessens, J. (1962) Man-made tornadoes. Nature. 193:13-14.

Eagan, R.C., P.V. Hobbs, and L.F. Radke (1974) Measurements of cloud condensation nuclei and cloud droplet size distributions in the vicinity of forest fires. J. Appl. Meteorol. 13:553-557.

Ensor, D.S., and M.J. Pilat (1971) Calculation of smoke plume opacity from particulate air pollutant properties. J. Air Pollut. Control Assoc. 21:496-501.

Evans, L.F., I.A. Weeks, A.J. Eccleston, and D.R. Packham (1977) Photochemical ozone in smoke from prescribed burning of forests. Environ. Sci. Technol. 11:896-900.

FEMA (1982) What the planner needs to know about fire ignition and spread. Chapter 3 in FEMA Attack Manual. Publication CPG 2-1A3. Washington, D.C.: Federal Emergency Management Agency.

Fitzgerald, J.W. (1973) Dependence of the supersaturation spectrum of CCN on aerosol size distribution and composition. J. Atmos. Sci. 30:628-634.

Glasstone, S. (1957) The Effects of Nuclear Weapons. Washington, D.C.: U.S. Department of Defense.

Glasstone, S., and P.J. Dolan (eds.) (1977) The Effects of Nuclear Weapons. Washington, D.C.: U.S. Department of Defense. 653 pp.

Goodale, T. (1971) The Ignition Hazard to Urban Interiors During Nuclear Attack due to Burning Curtain Fragments Transported by Blast. Report URS-7030-5. San Mateo, Calif.: URS Research Corp. 25 pp.

Greenfield, S.M. (1957) Rain scavenging of radioactive particulate matter from the atmosphere. J. Meteorol. 14:115-125.

Hegg, D.A., and P.V. Hobbs (1983) Preliminary measurements of the scavenging of sulfate and nitrate by clouds. Pages 79-89 in Precipitation Scavenging, Dry Deposition and Resuspension, edited by H.R. Pruppacher, R.G. Semonin, and W.G.N. Slinn. New York: Elsevier.

Hilado, C.J., and A.M. Machado (1978) Smoke studies with the Arapahoe chamber. J. Fire Flammability 9:240-244.

Hill, J.E. (1961) Problems of Fire in Nuclear Warfare. Report U 120869. Santa Monica, Calif.: RAND Corp.

Huschke, R.E. (1966) The Simultaneous Flammability of Wildlands and Fuels in the United States. RM-5073-TAB. Santa Monica, Calif.: RAND Corp.

Isaac, G.A., J.W. Strapp, H.A. Wiebe, W.R. Leaitch, J.B. Kerr, K.G. Arlauf, P.W. Summers, and J.I. MacPherson (1983) The role of cloud dynamics in redistributing pollutants and the implications for scavenging studies. Pages 1-13 in Precipitation Scavenging, Dry Deposition and Resuspension, edited by H.R. Pruppacher, R.G. Semonin, and W.G.N. Slinn. New York: Elsevier.

Ishikawa, E., and D.L. Swain (Transl.) (1981) Hiroshima and Nagasaki, The Physical, Medical and Social Effects of the Atomic Bombings. New York: Basic Books. 706 pp.

Issen, L.A. (1980) Single-Family Residential Fire and Live Loads Survey. Rep. NBSIR 80-2155. Gaithersburg, Md.: National Bureau of Standards. 176 pp.

Jaenicke, R. (1980) Atmospheric aerosols and global climate. J. Aerosol Sci. 11:577-588.

Jagoda, I.J., G. Prado, and J. Lahaye (1980) An experimental investigation into soot formation and distribution in polymer diffusion flames. Combust. Flame 37:261-274.

Janzen, J. (1980) Extinction of light by highly nonspherical strongly absorbing colloidal particles: Spectrophotometric determination of volume distributions for carbon blacks. Appl. Opt. 19:2977-2985.

Jaycor, (1980) Forest Environment (U). Report 5477 F. Washington, D.C.: Defense Nuclear Agency.

Kanury, A.M. (1976) The science and engineering of hostile fires. Fire Res. Abstr. Rev. 18:72-96.

Kemp, G. (1974) Nuclear Forces for Medium Powers. I. Targets and Weapons Systems. Adelphi Paper 106. London: International Institute of Strategic Studies. 41 pp.

Kent, J.H., and H.G. Wagner (1982) Soot measurements in laminar ethylene diffusion flames. Combust. Flame 47:53-65.

Kerker, M. (1969) The Scattering of Light and Other Electromagnetic Radiation. New York: Academic Press.

Kerr, J.W. (1971) Historic fire disasters. Fire Res. Abstr. Rev. 13:1-16.

Kerr, J.W., C.C. Buck, W.E. Cline, S. Martin, and W.D. Nelson (1971) Nuclear Weapons Effects in a Forest Environment--Thermal and Fire. Report N2:TR2-70. Washington, D.C.: Defense Nuclear Agency.

Kittelson, D.B., and D.F. Dolan (1980) Diesel exhaust aerosols. Pages 337-359 in Generation of Aerosols and Facilities for Exposure Experiments, edited by K. Willeke. Ann Arbor, Mich.: Ann Arbor Science.

Krinov, E.L. (1966) Giant Meteorites. New York: Pergamon Press. Pages 125-265 and 383-387.

Larson, D.A., and R.D. Small (1982a) Analysis of the Large Urban Fire Environment. II. Parametric Analysis and Model City Simulations. PSR Report 1210. Santa Monica, Calif.: Pacific Sierra Research Corp.

Larson, D.A., and R.D. Small (1982b) Analysis of the Large Urban Fire Environment. I. Theory. PSR Report 1210. Santa Monica, Calif.: Pacific Sierra Research Corp.

Leaitch, W.R., J.W. Strapp, H.A. Wiebe, and G.A. Isaac (1983) Measurements of scavenging and transformation of aerosol inside cumulus. Pages 53-69 in Precipitation Scavenging, Dry Deposition and Resuspension, edited by H.R. Pruppacher, R.G. Semonin, and W.G.N. Slinn. New York: Elsevier.

Leong, K.H., K.V. Beard, and H.T. Ochs III (1982) Laboratory measurements of particle capture by evaporating cloud drops. J. Atmos. Sci. 39:1130-1140.

Lewis, K.N. (1979) The prompt and delayed effects of nuclear war. Sci. Am. 241:35-47.

Liou, K.-N. (1980) An Introduction to Atmospheric Radiation. New York: Academic Press.

Lyman, H. (1918) Smoke from Minnesota forest fires. Mon. Weather Rev. 46:506-509.

Manins, P.C. (1984) Cloud heights and stratospheric injections resulting from a thermonuclear war. Submitted to Atmos. Environ.

Martin, S.G. (1974) The Role of Fire in Nuclear Warfare. Report URS 764 (DNA Report 2692F). San Mateo, Calif.: URS Research Corp. 166 pp.

Matson, M., S.R. Schneider, B. Aldridge, and B. Satchwell (1984) Fire detection using the NOAA-series satellites. NOAA Tech. Rep. NESDIS 7. Washington, D.C.: U.S. Department of Commerce.

McMahon, C.K. (1983) Characteristics of forest fuels, fires, and emissions. Paper presented at 76th Annual Meeting, Air Pollution Control Association, Atlanta, June 19-24.

McMahon, C.K., and P.W. Ryan (1976) Some chemical and physical characteristics of emissions from forest fires. Paper presented at the 69th Meeting, Air Pollution Control Association, Portland, June 27 to July 1.

Miller, C.F. (1962) Preliminary Evaluation of Fire Hazards from Nuclear Detonations. Memorandum Report, Project IMU-4021-302. Menlo Park, Calif.: Stanford Research Institute. 63 pp.

Miller, C.F., and J.W. Kerr (1965) Field Notes on World War II German Fire Experience. Report MU-5070. Menlo Park, Calif.: Stanford Research Institute. 69 pp.

Miller, R.K., M.E. Jenkins, and J.A. Keller (1970) Analysis of Four Models of the Nuclear-Caused Ignitions and Early Fires in Urban Areas. Report DC-FR-1210. Albuquerque, N.Mex.: Dikewood Corp.

Mordy, W.A. (1960) Discussion. Geophys. Monogr. 5. Am. Geophys. Union 5:399.

Morikawa, T. (1978) Evolution of soot and polycyclic aromatic hydrocarbons in combustion. Combust. Toxicol. 5:349-360.

Murgai, M.P. (1976) Natural Convection from Combustion Sources. New Delhi: Oxford and IBH Publishing.

Nakanishi, K., T. Kadoa, and H. Hiroyasu (1981) Effect of air velocity and temperature on the soot formation by combustion of a fuel droplet. Combust. Flame 40:247-262.

Office of Technology Assessment (1979) The Effects of Nuclear War. Washington, D.C.: Office of Technology Assessment.

Ogren, J.A. (1982) Deposition of particulate elemental carbon from the atmosphere. Pages 379-391 in Particulate Carbon: Atmospheric Life Cycle, edited by G.T. Wolff and R.L. Klimisch. New York: Plenum Press.

Ohlemiller, T.J., J. Bellan, and F. Rogers (1979) A model of smoldering combustion applied to flexible polyurethane foams. Combustion and Flame 36:197-215.

Oil and Gas Journal (1984) Statistics. Oil Gas J. 82:114.

O'Sullivan, E.F., and B.K. Ghosh (1973) The spectral transmission, 0.5-2.2 μm, of fire smokes. Pages 195-200 in Combustion Institute European Symposium 1973, edited by F.J. Weinberg. New York: Academic Press.

Packham, D.R., and R.G. Vines (1978) Properties of bushfire smoke: the reduction in visibility resulting from prescribed fires in forests. J. Air Pollut. Control Assoc. 28:790-795.

Pagni, P.J., and S. Bard (1979) Particulate volume fractions in diffusion flames. Seventeenth Symposium on Combustion. Pittsburgh, Pa.: The Combustion Institute. Pages 1017-1028.

Palmer, T.Y. (1976) Absorption by smoke particles of thermal radiation in large fires. J. Fire Flammability 7:460-469.

Palmer, T.Y. (1981) Large fire winds, gases and smoke. Atmos. Environ. 15:2079-2090.

Patterson, E.M., and C.K. McMahon (1983) Absorption Characteristics of Forest Fire Particulate Matter. Report FS-SE-2110-8. Atlanta, Ga.: U.S. Fire Service.

Patterson, E.M., B.T. Marshall, and K.A. Rahn (1982) Radiative properties of the Arctic aerosol. Atmos. Environ. 16:2967-2977.

Pluchino, A.B., S.S. Goldberg, J.M. Dowling, and C.M. Randall (1980) Refractive-index measurements of single micron-sized carbon particles. Appl. Opt. 19:3370-3372.

Plummer, F.G. (1912) Forest Fires: Their Causes, Extent and Effects, with a Summary of Recorded Destruction and Loss. U.S. For. Serv. Bull. 117:39 pp.

Powell, E.A., C.P. Bankston, R.A. Cassanova, and B.T. Zinn (1979) The effect of environmental temperature upon the physical characteristics of the smoke produced by burning wood and PVC samples. Fire Materials 3:15-22.

Prodi, F. (1983) The scavenging of submicron particles in mixed clouds: Physical mechanisms-laboratory measurements. Pages 505-516 in Precipitation Scavenging, Dry Deposition and Resuspension, edited by H.R. Pruppacher, R.G. Semonin, and W.G.N. Slinn. New York: Elsevier.

Pruppacher, H.R., and J.D. Klett (1978) Microphysics of Clouds and Precipitation. Boston: D. Reidel.

Quintiere, J.G. (1982) An assessment of correlations between laboratory and full-scale experiments for the FAA aircraft fire safety program. Part 1. Smoke. Report NBSIR 82-2508. Washington, D.C.: National Bureau of Standards. 33 pp.

Radke, L.F. (1983) Preliminary measurements of the size distribution of cloud interstitial aerosol. Pages 71-78 in Precipitation Scavenging, Dry Deposition and Resuspension, edited by H.R. Pruppacher, R.G. Semonin, and W.G.N. Slinn. New York: Elsevier.

Radke, L.F., J.L. Stith, D.A. Hegg, and P.V. Hobbs (1978) Airborne studies of particles and gases from forest fires. J. Air Pollut. Control Assoc. 28:30-34.

Radke, L.F., B. Benech, J. Dessens, M.W. Eltgroth, X. Henrion, P.V. Hobbs, and M. Ribon (1980a) Modifications of cloud microphysics by a 1000 MW source of heat and aerosol (the Meteotron project). Pages 119-126 in Proceedings of the Third WMO Scientific Conference on Weather Modification. Geneva: World Meteorological Organization.

Radke, L.F., P.V. Hobbs, and M.W. Eltgroth (1980b) Scavenging of aerosol particles by precipitation. J. Atmos. Sci. 19:715-722.

Radke, L.F., J.H. Lyons, D.A. Hegg, P.V. Hobbs, D.V. Sandberg, and D.E. Ward (1983) Airborne Monitoring and Smoke Characterization of Prescribed Fires on Forest Lands in Western Washington and Oregon. EPA Report 600/X-83-047. Washington, D.C.: U.S. Environmental Protection Agency. 104 pp.

Radke, L.F., J.H. Lyons, D.A. Hegg, and P.V. Hobbs (1984) Airborne monitoring and smoke characterization of prescribed fires on forest lands in western Washington and Oregon. Seattle: University of Washington.

Randhawa, J.S., and J.E. Van der Laan (1980) Lidar observations during dusty infrared Test-1. Appl. Opt. 19:2291-2297.

Rasbash, D.J., and D.D. Drysdale (1982) Fundamentals of smoke production. Fire Safety J. 5:77-86.

Rasbash, D.J., and B.T. Pratt (1979/80) Estimation of the smoke produced in fires. Fire Safety J. 2:23-37.

Roessler, D.M., and F.R. Faxvog (1979) Optoacoustic measurement of optical absorption in acetylene smoke. J. Opt. Soc. Am. 69:1699-1704.

Roessler, D.M., and F.R. Faxvog (1980) Optical properties of agglomerated acetylene smoke particles at 0.5145 μm and 10.6 μm wavelengths. J. Opt. Soc. Am. 70:230-235.

Roessler, D.M., D.-S.Y. Wang, and M. Kerker (1983) Optical absorption by randomly oriented carbon spheroids. Appl. Opt. 22:3648-3651.

Rosen, H., and T. Novakov (1983) Combustion-generated carbon particles in the Arctic atmosphere. Nature 306:768-770.

Rosenkilde, C.E., and F.J.D. Serduke (1983) Turbulent aspects of rainout. Pages 589-596 in Precipitation Scavenging, Dry Deposition and Resuspension, edited by H.R. Pruppacher, R.G. Semonin, and W.G.N. Slinn. New York: Elsevier.

Saito, F. (1974) Smoke generation from building materials. Fifteenth Symposium on Combustion. Pittsburgh, Pa.: The Combustion Institute. Pages 269-279.

Sandberg, D.V., J.M. Pierovich, D.G. Fox, and E.W. Ross (1979) Effects of Fire on Air. Technical Report WO-9. Washington, D.C.: Forest Service, U.S. Department of Agriculture. 40 pp.

Schroeder, M.J., and C.C. Chandler (1966) Monthly Fire Behavior Patterns. U.S. For. Serv. Res. Pap. PSW-112:15 pp.

Seader, J.D., and I.N. Einhorn (1976) Some physical, chemical, toxicological and physiological aspects of fire smokes. Sixteenth International Symposium. Pittsburgh, Pa.: The Combustion Institute. Pages 1423-1445.

Seader, J.D., and S.S. Ou (1977) Correlation of the smoking tendency of materials. Fire Res. 1:3-9.

Sehmel, G.A. (1980) Particle and gas dry deposition: A review. Atmos. Environ. 14:983-1011.

Seiler, W., and P.J. Crutzen (1980) Estimates of gross and net fluxes of carbon between the biosphere and the atmosphere from biomass burning. Climatic Change 2:207-247.

Short, N.M., P.D. Lowman, Jr., S.C. Freden, and W.A. Finch, Jr. (1976) Mission to earth: Landsat views the world. NASA SP-360. Washington, D.C.: National Aeronautics and Space Administration.

Shostakovitch, V.B. (1925) Forest conflagrations in Siberia. J. For. 23:365-371.

Slinn, W.G.N. (1977) Some approximations for the wet and dry removal of particles and gases from the atmosphere. Water Air Soil Pollut. 7:513-543.

Slinn, W.G.N. (1983) A potpourri of deposition and resuspension questions. Pages 1361-1416 in Precipitation Scavenging, Dry Deposition and Resuspension, edited by H.R. Pruppacher, R.G. Semonin, and W.G.N. Slinn. New York: Elsevier.

Small, R.D., and B.W. Bush (1984) Smoke production from multiple nuclear explosions in wildlands. Submitted to Science.

Smith, C.D., Jr. (1950) The widespread smoke layer from Canadian forest fires during late September 1950. Mon. Weather Rev. 78:180-184.

Tangren, C.D. (1982) Scattering coefficient and particulate matter concentration in forest fire smoke. J. Air Pollut. Control Assoc. 32(7):729-732.

Taylor, R.J., S.T. Evans, N.K. King, E.T. Stephens, D.R. Packham, and R.G. Vines (1973) Convective activity above a large-scale brushfire. J. Appl. Meteorol. 12:1144-1150.

Tewarson, A. (1982) Experimental Evaluation of Flammability Parameters of Polymeric Materials. Pages 97-153 in Flame Retardant Polymeric Material, Vol. 3, edited by M. Lewin, S.M. Atlas, and E.M. Pierce. New York: Plenum Press.

Tewarson, A., and J. Steciak (1982) Fire Ventilation. Technical Report. Washington, D.C.: U.S. Department of Commerce.

Tewarson, A., J.L. Lee, and R.F. Pion (1980) The influence of oxygen concentration on fuel parameters. Paper presented at the 18th International Symposium on Combustion, University of Waterloo, Canada, August 17-11.

Tien, C.L., D.G. Doornink, and D.A. Rafferty (1972) Attenuation of visible radiation by carbon smokes. Combust. Sci. Technol. 6:55-59.

Tomaselli, V.P., R. Rivera, D.C. Edewaard, and K.D. Möller (1981) Infrared optical constants of black powders determined from reflection measurements. Appl. Opt. 20:3961-3967.

Tsang, T.H., and J.R. Brock (1982) Effect of coagulation on extinction in an aerosol plume propagating in the atmosphere. Appl. Opt. 21:1588-1592.

Turco, R.P., O.B. Toon, T.P. Ackerman, J.B. Pollack, and C. Sagan (1983a) Nuclear winter: Global consequences of multiple nuclear explosions. Science 222:1283-1292.

Turco, R.P., O.B. Toon, T.P. Ackerman, J.B. Pollack, and C. Sagan (1983b) Global Atmospheric Consequences of Nuclear War. Interim Report. Marina del Rey, Calif.: R&D Associates. 144 pp.

Turco, R.P., O.B. Toon, R.C. Whitten, J.B. Pollack, and P. Hamill (1983c) The global cycle of particulate elemental carbon: A theoretical assessment. Pages 1337-1351 in Precipitation Scavenging, Dry Deposition and Resuspension, edited by H.R. Pruppacher, R.G. Semonin, and W.G.N. Slinn. New York: Elsevier.

Twitty, J.T., and J.A. Weinman (1971) Radiative properties of carbonaceous aerosols. J. Appl. Meteorol. 10:725-731.

Twomey, S. (1977) Atmospheric Aerosols. New York: Elsevier.

U.N. (1981) Demographic Yearbook, 1979. New York.

U.S. Department of Defense (1973) DCPA Attack Environment Manual, Chapter 3. Washington, D.C.

USDA (1972) A mathematical model for predicting fire spread in wildland fuels. U.S. For. Serv. Res. Note INT-115.

USDA (1981) Tree biomass--A state-of-the-art compilation. U.S. For. Serv. Tech. Rep. WO-33.

Uthe, E.E. (1981) Lidar evaluation of smoke and dust clouds. Appl. Opt. 20:1503-1510.

Valioulis, I.A., and E.J. List (1984) Collision efficiencies of diffusing spherical particles: Hydrodynamic, Van der Waals and electrostatic forces. Adv. Colloid. Interface Sci. 20:1-20.

Vervisch, P., D. Peuchberty, and T. Mohamed (1981) The spectral transmission of 0.4-4.5 μm of fire smokes. Combust. Flame, 41:179-186.

Volz, F.E. (1972) Infrared absorption by atmospheric aerosol substance. J. Geophys. Res. 77:1017-1031.

Wagner, H.Gg. (1981) Soot formation--An overview. Pages 1-29 in Particulate Carbon Formation During Combustion, edited by D.C. Siegla and G.W. Smith. New York: Plenum Press.

Wang, P.K., S.N. Grover, and H.R. Pruppacher (1978) On the effect of electric charges on the scavenging of aerosol particles by clouds and small raindrops. J. Atmos. Sci. 35:1735-1743.

Ward, D.E., R.M. Nelson, Jr., and D.F. Adams (1979) Forest fire smoke plume documentation. Paper 79-6-3 presented at the 72nd Annual Meeting, Air Pollution Control Association, Cincinnati, Ohio.

Watson, H.H. (1951) Alberta forest-fire smoke. Weather 6:253.

Watson, H.H. (1952) Alberta forest-fire smoke. Weather 7:128.

Wein, R.W., and D.A. MacLean (1983) The Role of Fire in Northern Circumpolar Ecosystems. New York: John Wiley and Sons. 322 pp.

Wexler, H. (1950) The great smoke pall--September 24-30, 1950. Weatherwise 3:129-134.

Wiersma, S.J., and S.B. Martin (1973) Evaluation of the Nuclear Fire Threat to Urban Areas. Report AD779-340. Menlo Park, Calif.: Stanford Research Institute. 131 pp.

Williams, D.W., J.S. Adams, J.J. Batten, G.F. Whitty, and G.T. Richardson (1970) Operation Euroka: An Australian Mass Fire Experiment. Report 386. Maribyrnor, Victoria, Australia: Defense Standards Laboratory.

Wolff, G.T., and R.L. Klimisch (eds.) (1982) Particulate Carbon: Atmospheric Life Cycle. New York: Plenum Press. 411 pp.

Woodie, W.L., D. Remetch, and R.D. Small (1983) Fire spread from tactical nuclear weapons in battlefield environments. PSR Note 566. Santa Monica, Calif.: Pacific Sierra Research Corp. 53 pp.

Wright, H.A., and A.W. Bailey (1982) Fire Ecology, United States and Southern Canada. New York: John Wiley and Sons.

APPENDIX 5-1: OBSERVATION OF PLUME HEIGHTS AND ASH TRANSPORT IN LARGE FIRES, by F.E. Fendell

Plume Heights

The altitude achieved by a plume over a maintained source of buoyancy depends largely on the strength of the source (heat released per unit time), the stratification and humidity of the ambient air, the strength of the crosswind (if any), and the size of the region of exothermicity. Rarely are all the desired inputs known for a single event.

As a reference, one of the more dramatic persistent plumes of the last quarter century was that associated with the creation of Surtsey off Iceland. An effective heat source estimated at 10^{11} J/s (with upflow at the base of roughly 120 m/s) was initiated at 7 A.M. on November 14, 1963 (the energy release rate was equivalent to about 250 kt every 3 h). By 10:30 A.M. the plume was at 3.5 km; by 3 P.M., at about 6.3 km; and by the next day, at over 9.3 km (i.e., to the height of the tropopause near Iceland). Vapor columns rose from neighboring sites on the sea to 2.5 km, and ash-laden steam burst upward to 0.6 km in a gigantic, ink-black column (Bourne, 1964; Thorarinsson and Vonnegut, 1964; Thorarinsson, 1966).

As another reference, the series of artificial convection experiments conducted at the Centre de Recherches Atmosphériques Henri Dessens, on the Lannemezan plateau in the French Pyrenees, entailed 105 fuel oil burners deployed in a three-arm spiral within a 140 m x 140 m square (the Meteotron). The heat release rate was about 10^9 J/s for 20 to 30 min (a total energy release of about 0.5 kt), and the plume reached 1 to 2 km (Benech, 1976; Church et al., 1980).

Plumes of most small-scale fires reach only a few kilometers into the troposphere. The black plume of a 10^{10} J/s oil fire that persisted for days near Long Beach, California, rose to 4 km (Hanna and Gifford, 1975). The convection column associated with the bombing of Leipzig in World War II, an event severe enough to give 15 m/s ground-level radial inflow at 4 km from the center and 34 m/s closer in, rose to only 3.9 km (Broido, 1960). The first thousand-bomber raid by the British in World War II (on Cologne, on May 30-31, 1942) produced a column of smoke that rose to 4.5 km (and hung as a huge pall at daybreak) (Barker, 1965). Taylor et al. (1973) reported a brushfire near Darwin River, Australia, on September 10, 1971, in which the ambient temperature fell almost linearly from 301 K at ground level to 268 K at 6 km. Whereas the plume rose to 3 to 4 km for a heat release rate of 10^{11} J/s, during a 10- to 15-min interval the plume advanced to 5.8 km when the heat release rate doubled. A small cloud above the plume was sucked down into it 10 min after this rapid additional ascent. However, the fuel loading for this case was about one-tenth that in portions of the American Pacific Northwest, which has the highest loadings in the continental United States. Thus one is motivated to examine severe burning events more closely.

Of the acreage burned in the United States annually, 95 percent comes from 2 to 3 percent of the total number of fires; these exceptional fires tend to occur in dry, hot, windy weather, can jump rivers and lakes, and decay only with wind shifts, the arrival of precipitation, and/or the exhaustion of fuel. Thirteen fire complexes in the recorded history of North America have each taken 4000 km^2 or more. Twelve thousand square kilometers were burned by the Maramichi and Maine fires of 1825, the North Carolina fire of 1898, and the Idaho and Montana fires of 1910; the Alaskan fires of 1957 consumed 20,000 km^2. Fire complexes in Michigan in 1871, in Wisconsin in 1894, and in Washington and Oregon in 1910 each burned 8000 km^2. Southern states lead the national fire statistics annually in both frequency and area burned; however, natural decomposition is slower in the North and fuel loads accumulate, so while the number of fires is fewer, with droughts come holocausts. As for extremes in spread rate, an 1887 Texas grass fire spread 26 km in 2 h, and crown fires propagating at 16 km/h have been recorded (Pyne, 1982).

At one time the August 1933 fire in Tillamook County, Oregon, was regarded as the most intense in recorded American experience. On August 24, 1933, hurricane-like winds arose, and 800 km^2 were burned in 20 h. The plume, which had reached 3 km (Holbrook, 1943), pierced an inversion, and the smoke column reached 11.1 to 12 km, near-tropopause-level altitude (Pyne, 1982).

In recent years, several events perhaps comparable in intensity to the Tillamook fire have been recorded. The Sundance fire in the northern Idaho area of Pack River and McCormick Creek advanced 14.5 km and burned 200 km^2 from 2 to 11 P.M. on September 1, 1967. The energy release rate is estimated at 5 x 10^{11} J/s, and the convection column reached 10.7 km, even though a 32- to 80-km/h wind was blowing (Anderson, 1968). The peak rate was achieved during saturation spotting in a valley somewhat sheltered from the wind.

A fire at an Air Force bombing range in North Carolina in 1971 was characterized by a crosswind of 32 km/h, a heat release rate of 1.2 x 10^{11} J/s, and a plume height of 4.6 km. A fire in the Sierra National Forest on July 16, 1961, burned 20 km^2 in 5 h, and a convective column rose to 6 to 9 km. The so-called Mack Lake fire in the Huron National Forest, Michigan, on May 5, 1980, burned 100 km^2 in 6 h; though the highly bent plume rose to only 4.6 km in the intense crosswind, the heat release rate has been estimated at 1.6 x 10^{11} J/s.

However, the highest free-burning-fire heat release rates are associated with firestorms, the exceptional heat-cyclone consequences of massive incendiary air raids on urban targets during World War II. The rareness of these events is evidenced by the fact that the U.S. Strategic Bombing Survey characterizes only four firestorms (Hamburg, Kassel, Darmstadt, and Dresden) arising from the 49 major German cities subjected to incendiary bombing (SPRI, 1975). No firestorm arose as a consequence of fifteen massive incendiary raids from March to June 1945 on Osaka, Kobe, Nagoya, Tokyo, or Yokohama, although the atomic bombing of Hiroshima produced a firestorm.

At Hamburg, during the raid on July 27-28, 1943, a cumulonimbus-cloud-like plume with an anvil top, of 3-km thickness, rose to 10 km (Ebert, 1963; Morton, 1970) in a near-adiabatic lapse rate in the lowest few kilometers of the troposphere; this altitude was ascribed by a meteorologist 6 km away, although Brunswig (1982, page 245) ascribes a height of only 7 km. Thick black smoke reached 6.9 km in half an hour after the onset of bombing; later-arriving crews reported severe turbulence, and some aircraft returned to base soot-covered (Middlebrook, 1981; Musgrove, 1981). Large black greasy raindrops fell along the outskirts of the fire (Caidin, 1960). Smoke and dust blotted out the sky for 30 h after the attack; the sun was not seen by Hamburg residents the next day (Rumpf, 1963).

Dresden was subjected to two massive raids on February 13-14, 1945, though stratocumulus clouds caused a total overcast above 3 km for most of the night, and strong winds persisted. In these raids, 12.4 km^2 were 75 percent destroyed, and an additional 4 km^2 were 25 percent destroyed, by fires that persisted 7 days and 8 nights. A firestorm occurred in a quarter circle of 2.2-km radius around the time of the raid. At daybreak on February 14, the city was obscured by a column of yellow-brown smoke filled with lifted flotsam; this column appeared particularly dark up to 4.8 km. Sooty ash showered downwind as far as 29 km for several days (Irving, 1965).

Smoke Obscuration

There are accounts of smoke so thick from Pacific Northwest forest fires that navigation on the Columbia River and other inland waterways was brought to a standstill in 1849 and 1868.

An instance of sun obscuration is given by the Peshtigo fires (October 8-9, 1871), in which 5000 km^2 were burned along both banks of the Green Bay. The sun was obscured for 320 km, and gloom persisted, even at noontime, for a week (Holbrook, 1943). Paper lofted from Michigan crossed Lake Huron and landed in Canada. On August 20, 1910, some 1750 separate fires in Idaho and Montana blew up and 12,000 km^2 were burned, such that the sun was blotted out (Holbrook, 1943). However, the time scale for reduced daytime visibility was days, not weeks.

References

Anderson, H.E. (1968) Sundance Fire: An Analysis of Fire Phenomena. Research Paper INT-56. Ogden, Utah: Intermountain Forest and Range Experiment Station, Forest Service, U.S. Dept. of Agriculture.

Barker, R. (1965) The Thousand Plan. London: Chatto and Windus.

Benech, B. (1976) Experimental study of an artificial convection plume initiated from the ground. J. Appl. Meteorol. 15:127-137.

Bourne, A.G. (1964) Birth of an island. Discovery 25 (April):16-19.

Broido, A. (1960) Mass fires following nuclear attack. Bull. Atmos. Sci. 16(10):409-413.

Brunswig, H. (1982) Feuerstrum über Hamburg. Stuttgart: Motorbuch Verlag.

Caidin, M. (1960) The Night Hamburg Died. New York: Ballantine.

Church, C.R., J.T. Snow, and J. Dessens (1980) Intense atmospheric vortices associated with a 1000 MW fire. Bull. Am. Meteorol. Soc. 61(7):682-694.

Ebert, C.H.V. (1963) The meteorological factor in the Hamburg fire storm. Weatherwise 16(2):70-75.

Hanna, S.R., and F.A. Gifford (1975) Meteorological effects of energy dissipation at large power parks. Bull. Am. Meteorol. Soc. 56(1):1069-1076.

Holbrook, S.H. (1943) Burning an Empire. New York: Macmillan.

Irving, D. (1965) The Destruction of Dresden. New York: Ballantine.

Middlebrook, M. (1971) The Battle of Hamburg. New York: Charles Scribner's Sons.

Morton, B.R. (1970) The physics of fire whirls. Fire Res. Abstr. Rev. 12:1-19.

Musgrove, G. (1981) Operation Gomorrah--The Hamburg Firestorm Raids. New York: Jane's.

Pyne, S.J. (1982) Fire in America--A Cultural History of Wildland and Rural Fire. Princeton, N.J.: Princeton University Press.

Rumpf, H. (1963) The Bombing of Germany. New York: Holt, Rinehart, and Winston.

Stockholm Peace Research Institute (1975) Incendiary Weapons. Cambridge, Mass.: MIT Press.

Taylor, R.J., S.T. Evans, N.K. King, E.T. Stephens, D.K. Packham, and R.G. Vines (1973) Convective activity over a large-scale bushfire. J. Appl. Meteorol. 12:1144-1150.

Thorarinsson, S. (1966) Surtsey, the New Island in the North Atlantic. Reykjavik, Iceland: Almenna Bokafelagio.

Thorarinsson, S., and B. Vonnegut (1964) Whirlwinds produced by the eruption of Surtsey volcano. Bull. Am. Meteorol. Soc. 45(8):440-444.

APPENDIX 5-2: WATER IN NUCLEAR CLOUDS

Clouds produced by nuclear explosions and by the fires they ignite can hold large quantities of water. The injection of this water into the upper air layers, and the consequences of the injection, are discussed in this appendix.

Explosion Clouds

Nuclear explosion clouds hold water that is vaporized and engulfed by the fireball. Surface bursts over deep water are expected to be relatively rare in a nuclear exchange and will be neglected (based on Pacific test data $\lesssim 3 \times 10^6$ tons of condensed water per megaton of yield are expected in the stabilized clouds (Gutmacher et al., 1983)). Subsurface ocean bursts do not generate high-altitude clouds (Glasstone and Dolan, 1977). Surface bursts over land can raise about 3×10^5 tons/Mt of soil to the stabilized cloud height. About an equal amount of groundwater and water of crystallization might be assumed. The fireball also entrains ambient water vapor as it rises through the lower troposphere. Adopting a fireball expansion rate such that $dR/dz \simeq 0.2$ (that is, the increase in the fireball radius is about one-fifth of the height traversed), and a U.S. Standard (1976) mid-latitude water vapor profile, the entrained water vapor could vary from $<1 \times 10^5$ to about 1×10^6 tons/Mt for a surface burst, depending on surface humidity. Accordingly, an average stabilized-cloud water content of 1×10^6 tons H_2O Mt is generous. The water concentration in stabilized nuclear clouds would be $\lesssim 1$ g/m^3, which is generally too small to cause precipitation but large enough to form an optically thick (ice) condensation cloud. As the nuclear cloud disperses, the ice particles would either settle out or evaporate. Air bursts above about 2 to 3 km would hold $<1 \times 10^5$ tons of H_2O per megaton.

The total water injected by explosion clouds in the baseline exchange would almost certainly be less than 6000 Tg. Most of the water would be deposited in the troposphere.

Fire Plumes

There are three sources of moisture for fire plumes: water of combustion, evaporated surface water, and entrained water vapor. Most combustible materials generate $\lesssim 1$ g-H_2O/g-burned. Thus, in the present baseline exchange, up to 8500 Tg of H_2O would be produced directly by fires, and could disperse with the plumes. Even if 1 cm of water were evaporated over the entire fire area in the baseline scenario, only 5000 Tg of additional water would enter the plumes; the actual amount would be much less, of course. Entrainment of ambient humidity into the plume, particularly at ground level where air can often be efficiently drawn into the fire, could add >1 g-H_2O/g-burned. At Hiroshima, a crude estimate suggests that about 10 g-H_2O/g-burned were entrained due to the high humidity at the time of the fire (R.P. Turco, private communication, 1984). However, most of this water fell as precipitation (the "black rain").

Due to condensation and precipitation, only a limited quantity of water can remain suspended in the fire plumes and be carried long distances. This quantity is assumed to be 5 g H_2O/g-burned, which consists primarily of moisture drawn into the fire near the ground. The total fire plume water injection in the baseline exchange may then be estimated as 40,000 Tg. The water is injected uniformly between 0 and 9 km (as is the smoke in the baseline case), or about 4000 Tg/km of altitude. Note that the injection represents primarily a redistribution of water vapor from the boundary layer into the free troposphere--as occurs during natural convection--because very little "new" water vapor is introduced by the combustion process.

The water concentration (condensate plus vapor) in the stabilized high-altitude plumes of large fires is expected to be about 1 g/m^3, based on the analysis of the water budget of a fire plume discussed above, air inflow rates obtained from plume theory, and direct measurements in fire cumulus cap clouds (L. Radke, private communication, 1984). The onset of condensation in the convective column of a fire may occur above the level expected for condensation in surface air lifted adiabatically, due to the added heat of combustion and the entrainment of dry ambient air aloft (Taylor et al., 1973). Low surface humidity, induced precipitation, and entrainment of dry air can all limit the water concentration in fire plumes.

The column abundances of water in fire clouds could be 1000 to 5000 g/m^2, compared to about 10,000 g/m^2 in natural cumulus clouds and 10 to 100 g/m^2 in cirrus clouds.

An upper limit to the water injection by fires in a nuclear conflict is in the vicinity of 500,000 Tg. This figure assumes that the initial fire plumes occupy a volume of 10^{17} m^3 (about one-tenth of the volume of the northern hemisphere mid-latitude troposphere), all of the air in the plumes originates in the surface layer and holds an average of 5 g H_2O/m^3, and no rainout occurs. Obviously, these circumstances are highly unlikely.

Water Perturbation

Table 5.2-1 gives the average ambient profile of water vapor at mid-latitudes. The global troposphere holds roughly 10^7 Tg of water vapor and the stratosphere, about 3000 Tg. If all of the water in nuclear explosion clouds were confined to the mid-latitude stratosphere, H_2O concentrations could increase by a factor of <10 there. Because the stratosphere normally is very dry, with a relative humidity of only 1 to 5 percent, and injected smoke and dust clouds can be heated by solar and infrared radiation, any condensed water would soon evaporate as the individual explosion clouds dispersed. A factor of 10 increase in stratospheric H_2O would affect ozone photochemistry and the infrared radiation balance of the stratosphere. The

TABLE 5.2-1 Ambient Atmospheric Water Vapor[a]

Altitude Interval (km)	Water Vapor Mixing Ratio[b] (ppmm)	Air Density[2] (kg/m^3)	Equivalent Global H_2O Mass in the Layer (Tg)	Cumulative H_2O up to the Top of Layer (Tg)
0-0.5	4686	1.225	$1.4x10^6$	$1.4x10^6$
0.5-1.5	3700	1.112	$2.1x10^6$	$3.5x10^6$
1.5-3.0	2843	1.007	$2.1x10^6$	$5.6x10^6$
3.0-5.0	1268	0.8194	$1.0x10^6$	$6.6x10^6$
5.0-7.0	554	0.6601	$3.7x10^5$	$7.0x10^6$
7.0-9.0	216	0.5258	$1.1x10^5$	$7.1x10^6$
9.0-11.	43.2	0.4135	$1.8x10^4$	$7.1x10^6$
11.-13.	11.3	0.3119	3500	$7.1x10^6$
13.-15	3.3	0.2279	750	$7.1x10^6$
15.-17.	3.3	0.1665	550	$7.1x10^6$

[a]Condensed water, which may reach concentrations of 10 g/m^3 (5×10^6 Tg globally in a 1-km-thick layer), is neglected.
[b]U.S. Standard Atmosphere (1976) Midlatitude Mean Model. The water vapor mixing ratio is given in parts per million by mass (ppmm). Local water vapor fluctuations typically exceed 10 percent.

photochemical effect of the H_2O, however, would probably not be any more important than the photochemical effect of the explosion-generated NO_x. The radiation perturbations are discussed below.

The fire plume water injection of about 4000 Tg/km up to 9 km is typically <1 percent of the ambient water vapor at any level in this height interval. The total fire H_2O injection is <0.5 percent of the global water vapor burden, and represents about 45 min of the normal global atmospheric water budget. The maximum perturbation could occur in the 7- to 9-km layer, where the average mid-latitude water vapor burden could increase by about 20 percent. If all of the fire water were put into this layer at northern mid-latitudes, the water burden would increase by about 0.20 g/m^3, or about 400 g/m^2. However, most of this water originates in lower regions of the atmosphere; the redistribution of water is likely to be less significant than an increase in the total water burden of the atmosphere.

The improbable "upper limit" water injections discussed in the previous section would lead to more substantial effects. Nevertheless, in view of the large ambient quantities of water vapor in the atmosphere, and the indirect water vapor perturbations to be discussed below, even the maximum credible water injections by fires could turn out to be of secondary interest.

CO_2 Perturbation

Carbon dioxide injections by nuclear fires are much less important than water injections. Because CO_2 is uniformly mixed throughout the troposphere and stratosphere (at about 340 parts per million by volume), the transfer of air between different altitude levels by nuclear explosions and fires has little effect on the CO_2 distribution. The global atmosphere holds about 3 x 10^6 Tg of CO_2. Nuclear fires could generate about 1 x 10^4 Tg of CO_2, roughly the amount produced in 1 year from fossil fuel combustion. Carbon dioxide is transparent in the visible spectrum, does not condense, and has only a limited infrared opacity (Liou, 1980). On the other hand, larger CO_2 perturbations might result from indirect disturbances in the global biospheric carbon cycle in the aftermath of a nuclear war (a subject that is not pursued in this report).

Effects of Water Injections

The water injected into the upper atmosphere with dust and smoke can have a number of important effects:

1. Modification of the photochemistry of ozone (see the previous discussion and Chapter 6).
2. Scavenging and washout of dust and smoke particles (see the discussion in Chapter 5).
3. Perturbation of the visible and infrared radiation balance by the condensed and vapor states of water.

During the first week after the start of a nuclear war, the localized explosion clouds and fire plumes could hold significant quantities of condensed water. The visible and infrared opacities of these clouds could be very large (>>1). Light levels below the clouds would be very low, particularly when heavy soot loadings are present. The infrared energy balance of the clouds would be complex, and some degree of thermal blanketing could result. Nevertheless, without solar insolation, the ground should still tend to cool. A strong greenhouse effect is not likely (at least in the case of smoke plumes) because solar absorption and heating would occur above most of the infrared opacity of the clouds (see Chapter 7).

In daylight, the smoke clouds would warm up rapidly, possibly inducing strong vertical and horizontal mixing of the cloud tops and edges, and perhaps causing some of the condensed water to evaporate. At night the clouds would cool by infrared emission, and subsidence might occur. The turbulence created by these heating and cooling cycles would be confined primarily to the upper cloud layers where precipitation is less probable. The major effect might therefore be to accelerate the dispersion of the smoke clouds. Some of the extended fire plumes would hold sufficient water to form thick cirrus anvils.

These cirrus could greatly increase the albedo above the smoke plumes, but would also hold in upwelling infrared radiation.

The large-scale advection and spreading of smoke clouds by self-induced heating has been studied on different spatial scales. Chen and Orville (1977) investigated cumulus-scale convection of carbon-black clouds. R. Haberle et al. (private communication, 1983) and M. MacCracken (private communication, 1984) simulated the motions of hemispherical-scale soot clouds. In each case, the same general behavior was predicted. The clouds tended to rise and spread horizontally at a faster rate than would be expected if only ambient air motions were acting. Direct observations of large sooty smoke clouds reveal a similar behavior (Davies, 1959).

Thus it is expected that some of the energy absorbed in the dust and smoke clouds would be converted into the kinetic energy of winds, which eventually dissipates as frictional heat.

Within about 2 weeks, the nuclear dust and smoke clouds would be sufficiently dispersed that their infrared opacities should be quite small (<1). The atmosphere could then approach the radiative regime analyzed by Turco et al. (1983a,b), Crutzen et al. (1984), and others, in which the infrared properties of the injected nuclear debris are less important than the visible properties.

Water vapor, particularly in the stratosphere, can affect the infrared radiation balance of the atmosphere. It has been estimated, for example, that a five-fold increase in stratospheric H_2O (with all other factors unchanged) would eventually lead to a 2°C surface warming (e.g., Manabe and Wetherald, 1967). However, in the nuclear perturbed atmosphere, even this modest effect is unlikely to occur, because the surface temperatures and infrared radiation fluxes of the lower atmosphere would already be greatly reduced.

Indirect Water Perturbations

Changes in surface air temperatures, winds, and atmospheric stability would disturb the "normal" hydrological cycle. Such disturbances could be more important than the primary water injections of the explosions and fires. Among the hydrological perturbations that might develop:

1. Increased low-level storminess and precipitation near ocean-continent margins, induced by exaggerated sea-land temperature contrasts.
2. Formation of widespread ground fogs over continents due to rapid radiative cooling of surface air.
3. Suppression of deep convection and upper-level precipitation caused by soot-induced heating of the upper troposphere.
4. Decrease in general cloudiness above several kilometers altitude as a result of warming and reduced relative humidity.
5. Reduction in the global water vapor burden associated with a general decrease in surface air temperatures.
6. Increase in water vapor concentrations above several kilometers altitude due to the enhanced moisture capacity of the heated air.

It is not likely that all of these effects would occur. A partial discussion of the possibilities is given in Chapter 7. Further research into these problems will be necessary to determine their importance.

References

Chen, C.-S., and H.D. Orville (1977) The effects of carbon black dust on cumulus-scale convection. J. Appl. Meteorol. 16:401-412.

Crutzen, P.J., C. Brühl, and I.E. Galbally (1984) Atmospheric effects from post-nuclear fires. Climatic Change, in press.

Davies, R.W. (1959) Large-scale diffusion from an oil fire. Pages 413-415 _in_ Atmospheric Diffusion and Air Pollution, edited by F.N. Frenkiel and P.A. Sheppard. New York: Academic Press.

Glasstone, S., and P.J. Dolan (eds.) (1977) The Effects of Nuclear Weapons. Washington, D.C.: U.S. Department of Defense. 653 pp.

Gutmacher, R.G., G.H. Higgins, and H.A. Tewes (1983) Total mass and concentration of particles in dust clouds. Rep. UCRL-14397. Livermore, Calif.: Lawrence Livermore Laboratory. 22 pp.

Liou, K.-N. (1980) An Introduction to Atmospheric Radiation. New York: Academic Press.

Manabe, S., and R.T. Wetherald (1967) Thermal equilibrium of the atmosphere with a given distribution of relative humidity. J. Atmos. Sci. 24:241-259.

Taylor, R.J., S.T. Evans, N.K. King, E.T. Stephens, D.R. Packham, and R.G. Vines (1973) Convective activity above a large-scale brushfire. J. Appl. Meteorol. 12:1144-1150.

Turco, R.P., O.B. Toon, T.P. Ackerman, J.B. Pollack, and C. Sagan (1983a) Nuclear winter: Global consequences of multiple nuclear explosions. Science 222:1283-1292.

Turco, R.P., O.B. Toon, T.P. Ackerman, J.B. Pollack, and C. Sagan (1983b) Global Atmospheric Consequences of Nuclear War. Interim Report. Marina del Rey, Calif.: R&D Associates. 144 pp.

U.S. Standard Atmosphere (1976) Washington, D.C.: U.S. Government Printing Office.

6 Chemistry

The production of nitric oxide by nuclear explosions and the production of soot and gaseous pollutants by fires ignited by nuclear explosions would pose chemical threats to the atmosphere in the postnuclear war period. This chapter first discusses the generation of the polluting substances and then assesses their impacts in the atmosphere.

GASEOUS EMISSIONS FROM NUCLEAR FIREBALLS AND NUCLEAR WAR FIRES

Nitric Oxide

Nuclear explosions produce nitric oxide (NO) by heating air to very high temperatures both in the interior of the fireball and in the accompanying shock wave. At temperatures above about 2000 K the equilibrium

$$N_2 + O_2 \rightleftharpoons 2NO$$

is rapidly established, the amount of NO increasing with increasing temperature. As hot air containing large amounts of NO is cooled, the above equilibrium is maintained until a temperature is reached where the rates of the reactions maintaining the equilibrium become slow in comparison with the cooling rate. For cooling times of seconds to milliseconds, the NO concentration "freezes" (becomes fixed) at temperatures between 1700 K and 2500 K, corresponding to NO concentrations of 0.3 and 2.0 percent by volume, respectively.

There have been numerous estimates of the amount of NO produced per megaton of explosion energy, and these have been reviewed by Gilmore (1975). The spherical shock wave is estimated to produce 0.8×10^{32} NO molecules per megaton of explosive yield, as a result of the rapid heating of air in the shock front followed by rapid cooling due to expansion and radiative emission.

The shock wave calculation of NO production does not take into account the fact that air remaining within the fireball center contains approximately one-sixth of the initial explosion energy. This air cools on a time scale of several seconds by further radiative emission, entrainment of cold air, and expansion as it rises to higher altitudes. These mechanisms are sufficiently complex that one can only estimate upper and lower limits to the quantity of NO finally produced.

A lower limit to the amount of NO finally produced may be obtained by assuming that all of the shock-heated air is entrained by the fireball and (again) heated to a temperature high enough to reach equilibrium. This is possible since the thickness of the "shell" of shock-heated air containing NO is smaller than the radius of the fireball. To minimize the cooling rate, and thus the freeze-out temperature, it is assumed that this air mass cools only by adiabatic expansion as the fireball rises and by using a minimum rise velocity. The resulting lower limit to NO production is 0.4×10^{32} molecules per megaton.

Since the interior of the fireball is much hotter than the surrounding shock-heated air, it will rise much faster and possibly pierce the shell of shock-heated air to mix with the cold, undisturbed air above it. Thus an upper limit to NO production may be obtained by assuming that none of the 0.8×10^{32} NO molecules per megaton produced in the shock wave are entrained by the hot fireball and that the interior is cooled totally by entrainment of cold, undisturbed air to produce additional NO. The upper limit to total NO production is then estimated to be 1.5×10^{32} molecules per megaton. One can make strong arguments that both the lower and the upper limits are extremely unlikely. For the purposes of this assessment, a NO yield of 1.0×10^{32} molecules per megaton (0.005 Tg/Mt) is assumed.

It should be emphasized that the emission factor for nitrogen oxides produced in nuclear explosions is based wholly on theoretical considerations and that there has not yet been any attempt at experimental verification of the amounts produced. Sedlacek et al. (1983), analyzing samples for HNO_3 in the stratosphere, infer that the Chinese 4-Mt nuclear device of 1976 produced about 10 times the amount of NO as expected from the theoretical calculations discussed above. The discrepancy remains unexplained. Table 6.1 gives the calculated total amounts of NO injected by the two scenarios of this study as well as amounts used in other studies.

Fire Emissions

Uncontrolled fires result in incomplete combustion with emissions of copious quantities of both particulate and gaseous matter. The particulate emissions and their effects on the physical properties of the atmosphere are dealt with in other sections of this report. Among the gaseous emissions from fires are carbon monoxide, nitrogen oxides, and a large number of hydrocarbons and other organic compounds. These compounds, together with sunlight, are the necessary ingredients for photochemical smog formation.

TABLE 6.1 Recent Estimates of Maximal Ozone Depletion Resulting from a Nuclear War

Scenario	Yield (Mt)	NO (10^{32} molecules) Below 12 km	NO (10^{32} molecules) Above 12 km	Maximum Ozone Depletion (percent)	Note
Baseline	6,500	2,665	3,835	17	a
Excursion	8,500	2,665	5,835	43	b
Chang Case A	10,600	560	6,540	51	c
Chang Case B	5,300	280	3,270	32	d
Chang Case C	5,670	0	3,800	42	e
Chang Case D	4,930	560	2,740	16	f
Chang Case E	6,720	180	4,340	39	g
Chang Case F	3,890	390	2,220	20	h
Ambio	5,740	4,510	1,230	~0	i
Ambio Excursion	10,000	1,375	8,625	65 (45°N)	j
Turco et al. (1983)	10,000	1,200	8,400	50	k

[a]No weapons larger than 1.5 Mt. See Chapter 3 for details.
[b]Baseline scenario plus 100 weapons of 20-Mt yield.
[c]All strategic weapons in the United States and USSR arsenals successfully detonated.
[d]Half of the weapons of each type in the strategic arsenals of the United States and USSR.
[e]All weapons with individual yields greater than 0.8 Mt in the strategic arsenals of the United States and USSR.
[f]All weapons with individual yields less than or equal to 0.8 Mt in the strategic arsenals of the United States and USSR.
[g]All weapons in the Soviet strategic arsenal.
[h]All weapons in the U.S. strategic arsenal.
[i]When the troposphere is included, the *Ambio* scenario actually results in a slight ozone increase. The Chang model also gives this result for the *Ambio* scenario.
[j]The *Ambio* excursion scenario consists of 5000 1-Mt detonations plus 500 10-Mt detonations and is identical to the NRC (1975) scenario.
[k]The blocking of sunlight by nuclear dust and soot was accounted for, but the resulting heating of the stratosphere was not.

Carbon Monoxide

Carbon monoxide (CO) is the most abundant air pollutant from fires. The emission factor may be quite high, depending on the degree of aeration. For example, Sandberg et al. (1975) measured CO emissions in the range of 25 to 40 percent in very low intensity laboratory fires, and Ryan and McMahon (1976) state that CO emissions may approach 25 percent for smoldering fires in damp fuels.* Emissions from prescribed forest fires fall in the range of 1 to 25 percent according to Tangren et al. (1976). A review by Chi et al. (1979) recommends an emission factor of 5.6 ± 1.6 percent for prescribed fires where the indicated error is the 95 percent confidence interval of the mean. The committee has adopted an emission factor for CO of 5 percent for its calculations. For the baseline scenario in which 4500 Tg of fuel is consumed by fire, the CO emission is 225 Tg. Mixed uniformly throughout half of the northern hemisphere troposphere, the concentration of CO is increased from the present level of about 100 ppbv (parts per billion by volume) to about 300 ppbv.

Hydrocarbons

Hydrocarbons are an extremely diverse class of organic compounds consisting only of carbon and hydrogen. They include aliphatic hydrocarbons (alkanes) such as methane, ethane, and propane; olefins (alkenes) such as ethylene and propylene; alkynes such as acetylene; and aromatic compounds such as benzene, toluene, and the xylenes. In addition to the hydrocarbons, numerous oxygen-containing compounds such as alcohols, ethers, aldehydes, ketones, and carboxylic acids have been identified in fire emissions. In fact, more than 200 individual compounds have been identified in forest fire emissions, and considering the results of recent studies of cigarette smoke, it is likely that the actual number of compounds emitted is in the many thousands. Figures 6.1 and 6.2 are chromatograms of air collected above slash burning in the tropical forests of Brazil (Greenberg et al., 1984). These chromatograms illustrate the numerous compounds typically found in fire emissions.

Total hydrocarbon emissions have been reported to be in the range of 0.2 to 3.2 percent in laboratory fires and 1.4 to 5.4 percent in a limited number of field fires (McMahon, 1983). In recent measurements of biomass burning in Brazil (Greenberg et al., 1984), emission factors of 2.0 percent for grassland fires and 2.7 percent (expressed as percent of carbon dioxide by volume) for forest fires were obtained. Methane made up 35 percent of the grassland emissions and 50 percent of the forest emissions.

As an estimate for nuclear war fires, the committee has adopted an emission factor of 2 percent (based on weight of fuel burned) for total hydrocarbons and further has assumed that half of these emissions are due to methane, which is relatively less reactive than the higher hydrocarbons. For the baseline scenario, this results in a total emission of 45 Tg of methane and 45 Tg of other hydrocarbons. The methane emission results in an increase in its concentration at mid-latitudes of 70 ppbv. This represents only a minor increase in the ambient methane concentration of 1650 ppbv. As a result, the methane

---

*Unless otherwise noted, all emission factors quoted refer to the mass of a particular chemical species produced per mass of fuel consumed.

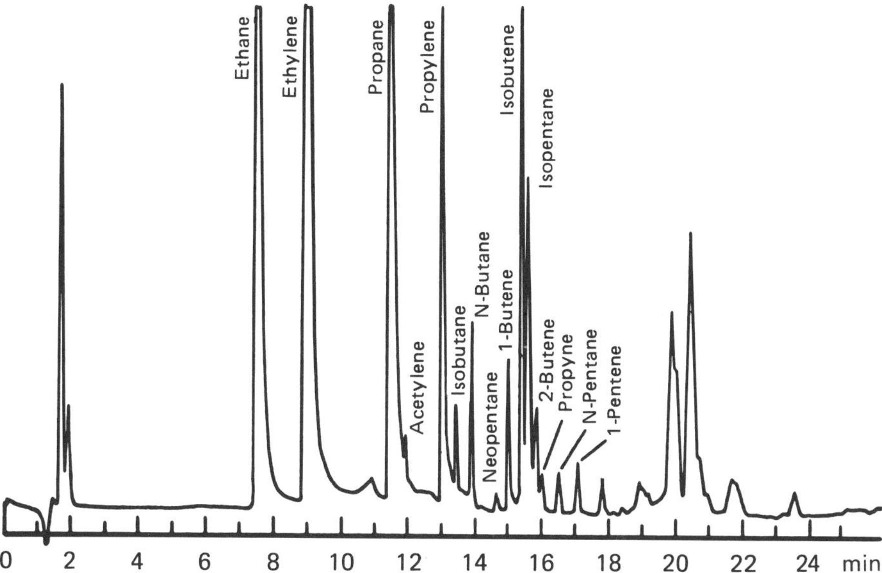

FIGURE 6.1 Light hydrocarbon chromatogram of air collected above slash burning in the tropical forests of Brazil (Greenberg et al., 1984).

input may be ignored in the calculations of tropospheric photochemistry.

To represent the nonmethane hydrocarbons, the committee has chosen to distribute the emissions as 25 percent ethane, 25 percent propane, and 50 percent ethylene, as these are the major compounds observed in fire emissions, approximately 50 percent of which are alkenes.

Oxides of Nitrogen

Data for nitrogen oxide emissions in forest and urban fires are still limited. The flame temperatures are generally not high enough in isolated fires to produce nitric oxide directly from air. The fixed nitrogen in the urban and forest fuels represents another fire-produced source of oxides of nitrogen. For example, for ponderosa pine the nitrogen content ranges from 0.1 percent in boles to 1 percent in growing needles (Tangren et al., 1976). The Environmental Protection Agency has assigned an emission factor of 0.2 percent (as nitrogen dioxide) based on laboratory burning of landscape refuse (McMahon, 1983). Ward et al. (1982) recently reported a value of 0.18 percent from burning forest materials in a field study. Another study (DeAngelis et al., 1980) found a nitrogen dioxide emission factor of 0.19 percent for the burning of wood in fireplaces. Burning of wood, bark, and limbs at temperatures below 1000°C gave an average emission factor of 0.15 percent, compared to 0.75 percent for pine needles and other forest foliage (Clements and McMahon, 1980). Considering that lumber makes up most of the urban fuels, the committee has adopted a conservative emission factor of 0.15 percent. This results in a total

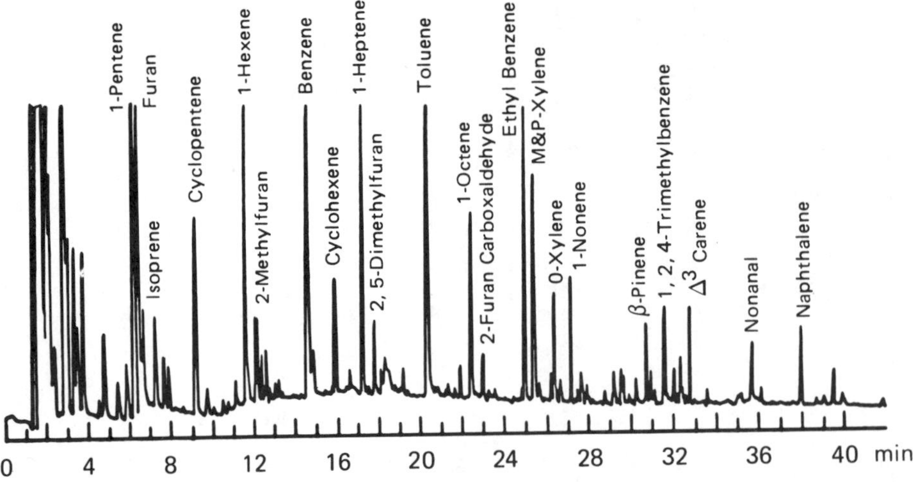

FIGURE 6.2 Heavy hydrocarbon chromatogram of air collected above slash burning in the tropical forests of Brazil (Greenberg et al., 1984).

emission to the atmosphere of 6.8 Tg of nitrogen dioxide (equivalent to 4.4 Tg of nitric oxide) from fires for the baseline scenario.

EFFECTS OF EMISSIONS

Ozone Shield Reduction

The first perceived threat of stratospheric ozone by pollutants implicated the oxides of nitrogen (NO and NO_2, known collectively as NO_x). At that time, the early 1970s, it was the prospect of supersonic flight that caused concern (see, e.g., NRC, 1973). Threats to the ozone layer from emissions of chlorofluorocarbons and from increases in nitrous oxide (N_2O) concentrations (caused by the increased application of nitrogen fertilizers) have been recognized and assessed (see, e.g., NRC, 1982). The problem of ozone reduction by N_2O increases is in essence the same as that of reduction by adding NO_x, since N_2O is converted to NO in the stratosphere. In 1975 the NRC conducted a workshop for the purpose of studying effects of large-scale nuclear detonations. Of all of the aspects addressed, that concerning the effects of NO_x injection received the most detailed treatment because of the recent awareness brought about by the SST studies (Crutzen, 1971; Johnston, 1971) and the work of Foley and Ruderman (1973), who pointed out that the NO_x produced in the fireballs of nuclear weapons should lead to ozone reduction (see also Johnston et al., 1973). Recently, estimates have been made of ozone reductions from NO_x injections for various nuclear war scenarios (Chang and Wuebbles, 1982; Crutzen and Birks, 1982). The results of all of these studies and their respective scenarios are summarized in Table 6.1.

The list of chemical reactions thought to describe the behavior of ozone in the stratosphere is long and imposing. The interactions of the various atoms and molecules among themselves and with sunlight and their further dependency upon atmospheric transport make up a very complicated system. Though much is known about this system and the ability to model it has increased considerably in the last decade, much uncertainty still remains attendant to the application of the models to such drastic perturbations as those in the baseline scenario. However, there is now a large body of evidence that concentrations of ozone in the present stratosphere are principally controlled by NO_x from natural sources. For this reason alone, it is expected that a large perturbation in the stratospheric burden of NO_x, particularly in the upper regions of the stratosphere, would result in a large decrease in the ozone column.

The committee attempts here to give only a brief explanation of the manner in which NO_x causes ozone reduction in the baseline and excursion cases. (For a thorough review of the chemistry of stratospheric ozone, the reader is referred to Logan et al. (1978). An update is available in the appendixes by Wofsy and Logan and by Anderson in NRC (1982).) Figure 6.3 shows a "normal" ozone concentration vertical profile and the altitude ranges into which the NO_x would be deposited in the 6500-Mt baseline scenario and the 8500-Mt excursion scenario. The ozone concentrations are controlled by balances of production and loss reactions and transport. There are several sets of photochemical reactions, some of which form cycles that can explain much of the observed behavior of ozone. These cycles include catalytic destruction of odd oxygen (O_3 and O atoms) by the oxides of nitrogen, the odd hydrogen radicals (HO and HO_2), and the chlorine radicals (Cl and ClO). The pertinent cycle for ozone destruction by NO_x is the set of reactions:

$$
\begin{array}{l}
NO + O_3 \rightarrow NO_2 + O_2 \\
NO_2 + O \rightarrow NO + O_2 \\
\hline
O_3 + O \rightarrow 2O_2
\end{array}
$$

At mid-latitudes in the normal atmosphere, this reaction cycle provides the principal means of odd oxygen destruction above about 23 km. Although the cycle also provides most of the chemical loss of odd oxygen at lower altitudes, the rate of the NO_2 + O reaction (which limits the rate of the cycle) slows in relation to the rate of transport as the altitude decreases below about 23 km. The amount of ozone reduction caused by injection of NO into the stratosphere depends on the amounts of NO and their distribution with altitude, which in the case of a nuclear bomb depend upon the yield and height of burst. Figure 4.3 shows the distribution of nuclear cloud tops and bottoms used to calculate the distributions of injected NO in model calculations of ozone reduction. Thus the estimate of the ozone reduction that would result from a nuclear war depends on the yield, type of burst, and latitude, for each weapon of the scenario used. For the baseline scenario, concentrations of NO_x would be greatly enhanced in the lower stratosphere up to about 19 km.

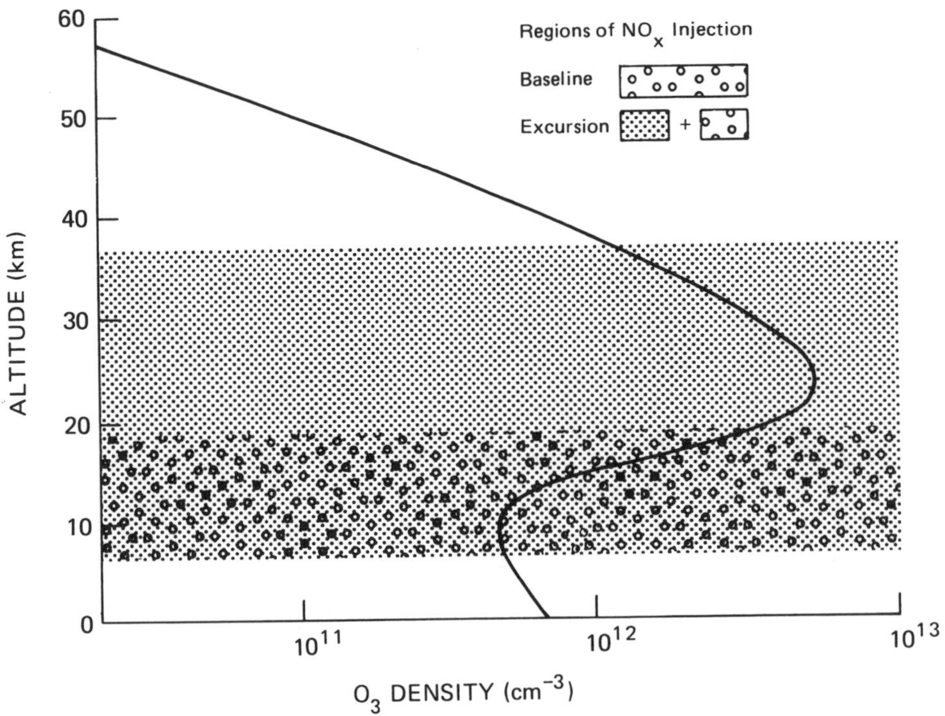

FIGURE 6.3 Concentration (solid line) of ozone in the unperturbed atmosphere at regions of NO_x injection.

The ozone reduction at these elevations would occur because the rate of the catalytic cycle shown above would be enhanced relative to removal by transport. The ozone destruction rate would also increase as the oxides of nitrogen mixed upward, where the ozone concentrations are higher and the photochemical reaction rates are faster. It is a technically noteworthy point that for this massive low-level injection of NO_x, below the level of the ozone maximum at 25 km, the overlying ozone would prevent compensating odd oxygen production resulting from photolysis of O_2 at the lower elevations.

The Lawrence Livermore National Laboratory (LLNL) one-dimensional eddy diffusion and chemical reaction model was used to estimate the amounts of ozone reduction with time corresponding to the present two scenarios (J.E. Penner and P.S. Connell, Lawrence Livermore National Laboratory, private communication, 1984). The results are shown in Figure 6.4, which illustrates that for the baseline case the maximum ozone reduction of 17 percent (average over the northern hemisphere) would be reached 1 year after the war and recovery to one-half of the peak reduction would require an additional 2 years. The relatively slow development of the ozone minimum reflects primarily the slow upward transport of NO_x to regions where the odd oxygen destruction rates are greater.

The 8500-Mt excursion scenario would place additional large amounts of NO_x at elevations up to about 37 km due to the use of much larger

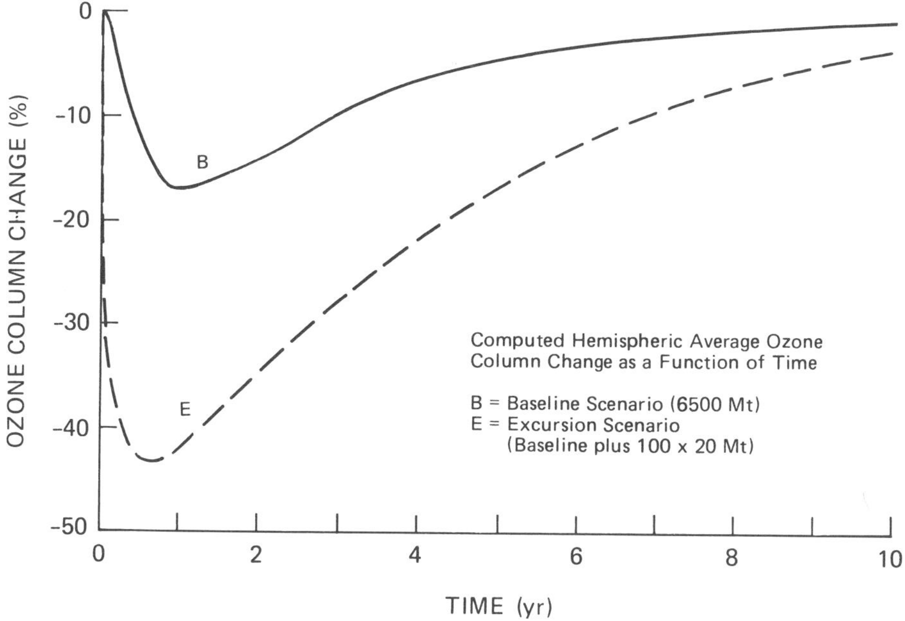

FIGURE 6.4 Hemispherically averaged percent ozone depletion estimated in a one-dimensional eddy diffusion and chemical reaction model (J.E. Penner and P.S. Connell, LLNL, private communication, 1984).

weapons. This would cause very rapid reduction of ozone in the region where its concentrations are the highest. This is reflected in the shorter time to achieve maximum reduction, namely about 8 months to reach 43 percent reduction. Recovery to one-half that value would occur after 4 years.

The complex set of chemical reactions that control stratospheric ozone concentrations constitutes a system in which the dependency of ozone reductions amounts on NO injection amounts is somewhat nonlinear. These effects were discussed in the NRC (1975) report (see particularly Figure 1.9). Though there are differences in details between the model used then and the present model, the plot is still approximately applicable to the present model for the purpose of rough guidance.

The results of this study are consistent with those of other studies using the LLNL model. Comparison of these results with those reported by Chang and Wuebbles (1982) shows the same shapes for the ozone versus time curves. Table 6.1 presents a comparison of the present scenarios, NO_x injections, and maximum ozone (O_3) reductions with those of Chang and Wuebbles and of the *Ambio* scenario. The details of time to maximum reduction, the value of maximum ozone reduction, and time to recover to one-half the maximum depletion are scenario-dependent. That is, the amount and distribution of NO_x and thus the model-derived details depend on the numbers and yields of individual weapons. (The results for the excursion scenario are somewhat similar to those reported in NRC (1975). However, there are substantial differences in several of the reaction rate parameters as well as yields and numbers of weapons.)

The Ambio scenario gave no ozone reduction and is consistent with the results of the present study for the reason that the preponderance of the NO was injected below 12-km altitude. There the ozone destruction by the NO_x cycle is offset by the creation of ozone by the smog cycle (see next section).

The calculated ozone changes discussed above were obtained with the use of models in which the stratospheric transport properties are those that represent the present (unperturbed) atmosphere. The perturbation of the atmosphere by smoke and dust could affect the circulation of the stratosphere and thus provide a circumstance different from that upon which the present ozone calculations were based. As discussed in Chapter 7, the committee has no sound qualitative notions as to how stratospheric circulation would be altered. In the absence of further information, the committee believes that the use of the unperturbed atmospheric transport characteristics provides the best basis for assessing the ozone reductions caused by NO injections from nuclear bursts.

Since the model used in this study considers transport only in the vertical dimension, it cannot provide an estimate of the amounts of NO_x transported into the southern hemisphere. The ability of the atmosphere to transport trace substances across the equator in the stratosphere was demonstrated by many observations of radioactive debris from nuclear weapons testing in the atmosphere. The nature of this phenomenon was delineated by Mahlman and Moxim (1978) using a general circulation model. Their study, using a single mid-latitude tracer injection, showed that the maximum burden in the southern hemisphere occurred about 9 months after the injection and was less than 10 percent of the initial amount injected. Crutzen and Birks (1982) calculated southern hemisphere ozone reduction to be of the order of 15 percent occurring after the injection of somewhat higher amounts of NO_x than in the excursion case.

Ozone Holes and Effects of NO_2 Radiation Absorption

Luther (1983) has studied short-term chemical and radiative effects of injections of NO into the stratosphere by nuclear weapons. The particular problem he addresses is the "ozone hole." Rapid heating of portions of the stratosphere containing high concentrations of NO_2, with subsequent mixing throughout the heated and destabilized volume, causes the ozone hole, which is a large reduction in the ozone column abundance distributed over most of the vertical extent of the stratosphere, but confined laterally. Ozone holes would permit a very large increase in irradiance of ultraviolet light at the top of the troposphere, which, in the absence of smoke or clouds, would result in life-damaging effects at the surface. Luther's study assumed that the cloud remained cylindrical throughout the depths of the stratosphere and that horizontal mixing could be represented by eddy diffusion.

These assumptions are probably not realistic, since the "filling" of the holes by shear in the vertical is likely to be rapid and effective. Thus, it is considered that the ozone holes would exist for no more than a few hours and their effects would be less severe than those from global-scale reduction.

Effects on Ozone of Past Nuclear Weapons Tests

In accordance with the committee's estimates, the approximately 300 Mt of total bomb yield in multimegaton atmospheric bursts by the United States and USSR in 1961 and 1962 introduced about 3 x 10^{34} additional molecules of nitric oxide into the stratosphere. Thus one might ask whether these tests resulted in a depletion of the ozone layer. Using a one-dimensional model, Chang et al. (1979) estimated that these nuclear weapons tests should have resulted in a maximum ozone column depletion in the northern hemisphere of about 4 percent in 1963. Analysis of the ground ozone observational data by Johnston et al. (1973) showed a decrease of 2.2 percent for the period 1960-1962 followed by an increase of 4.4 percent in 1963-1970. Although these data are consistent with the magnitude of the ozone depletion expected, by no means is a cause and effect relationship established. Angell and Korshover (1973) attribute these observed ozone column changes to meteorological factors. The ozone decrease began before most of the large weapons had been detonated and persisted for too long a period to be totally attributed to recovery from bomb-induced ozone depletion. Unfortunately, because of the large scatter in the ground-based ozone observational data and our lack of understanding of all of the natural causes of ozone fluctuations, one cannot draw definite conclusions about the effects of nuclear explosions on stratospheric ozone on the basis of previous tests of nuclear weapons in the atmosphere.

Uncertainty in Model Results

Normally, a scientific study using a model to "predict" a result should be accompanied by an analysis of uncertainties ending with a set of error limits on that result. The assessment of global effects of perturbing trace substances on stratospheric ozone has caused much effort to be expended in attempting to estimate error limits on calculated ozone reductions. Yet after more that a decade of experience in this exercise the most recent assessment by the NRC (1984a) states, "The detailed treatments often leave the wrong impression that the actual sources of uncertainty are well defined. . . . [O]nly a qualitative statement of uncertainty is made here." The perturbation of the stratosphere by NO and smoke emissions from a large-scale nuclear war is likely to be so large that effects not considered could well play an important role. Certainly, the models used in the present assessment were not constructed to handle such perturbations. Further, the present problem is complicated by the many injections of NO in the vicinity of the tropopause by low-yield weapons. The estimated ozone depletion for the baseline case is quite sensitive to the height of the tropopause relevent to the particular bursts. As discussed in Chapter 4, there is uncertainty associated with the estimates of cloud tops and bottoms. All of these factors combine to make of any rational estimate of error limits in ozone reduction a virtual impossibility. The numbers calculated here, though given to two figures, should be viewed as plausible values that are based upon the best methods available to the committee.

Tropospheric Composition Changes

Because the troposphere is in direct contact with the biosphere, it is especially important to understand the chemical changes that would take place in this region of the atmosphere following a nuclear war. The many fires ignited by the nuclear explosions would inject large quantities of carbon monoxide, hydrocarbons, and many other organic compounds into the atmosphere. Both fires and the nuclear explosions themselves would produce large quantities of oxides of nitrogen. In the presence of sunlight, these compounds react to form strong oxidants, particularly ozone and organic peroxides such as peroxyacetyl nitrate (PAN). PAN and related compounds have strong phytotoxic effects. Ozone, while being necessary in the stratosphere to serve as a shield against solar ultraviolet radiation, is considered undesirable at ground level because of its toxic effects on both plants and animals.

Whether or not a dense photochemical smog with high oxidant concentrations would form in the wake of a nuclear war is difficult to evaluate for several reasons. Perhaps the largest uncertainties are associated with (1) the extent and duration of the darkening caused by the smoke and dust, and (2) changes in tropospheric dynamics and precipitation rates, which in turn affect the lifetimes of the relevant chemical species. The generalized mechanism of photochemical smog formation includes the critical reaction sequence

$$ROO + NO \rightarrow NO_2 + RO$$
$$NO_2 + h\nu \rightarrow NO + O$$
$$O + O_2 + M \rightarrow O_3 + M$$

where R can be a hydrogen atom or any organic radical and M is any molecule. This sequence of reactions requires sunlight (photon, $h\nu$) and oxides of nitrogen (NO and NO_2). Sunlight is also necessary to the formation of the hydroxyl radical, OH, as follows,

$$O_3 + h\nu \rightarrow O_2 + O(^1D_2)$$
$$O(^1D_2) + H_2O \rightarrow 2\ OH$$

where $O(^1D_2)$ is an electronically excited oxygen atom. The OH radical is an important initiator of chain reactions in the atmosphere via reactions such as

$$CO + OH \rightarrow CO_2 + H$$

followed by

$$H + O_2 + M \rightarrow HOO + M$$

and

$$RH + OH \rightarrow H_2O + R$$

which is followed by

$$R + O_2 + M \rightarrow ROO + M$$

J.E. Penner and P.S. Connell (Lawrence Livermore National Laboratory, private communication, 1983) have investigated the tropospheric composition changes associated with the baseline scenario using a one-dimensional model of tropospheric photochemistry. Because most of the oxides of nitrogen in the troposphere are removed in this model by natural processes of dry deposition and rainout during the first few weeks, while the sunlight is greatly attenuated by suspended smoke and dust, the average concentration of ozone in the troposphere increases by less than a factor of 2. After several more weeks, the ozone concentration is expected to have decreased to near-ambient levels as the many chemical pollutants are removed from the atmosphere.

The high loading of particulate matter in the troposphere may be significant not only in blocking sunlight, but also in promoting heterogeneous reactions. Assuming all smoke particles are perfect spheres of radius 0.05 μm with a density of 1 g/cm^3, the specific surface area is 60 m^2/g. If the 200 Tg of smoke aerosol of the baseline case is uniformly distributed with a constant mixing ratio (aerosol particles/molecules of air), then every atmospheric molecule collides with a particle on the average about 4 times every second. This collision lifetime is shorter that the lifetime of many highly reactive atmospheric species. Birks and Staehelin (1984) have investigated the possible role of reactions on particulate surfaces in further reducing tropospheric oxidant concentrations. They found for the baseline case that oxidant formation in the troposphere is significantly inhibited when the efficiencies (γ) of reaction upon collision with aerosol surfaces exceed 10^{-6} for O_3, 10^{-1} for OH and/or 10^{-2} for HO_2. The variations in values of γ that result in significant reduction in oxidant formation simply reflect the relative lifetimes of oxidant species in the atmosphere. Whereas a small value of γ_{O_3} is required for ozone, a relatively long-lived species, a value of $\gamma_{OH} \geq 10^{-1}$ is required for hydroxyl radicals, which have a very short atmospheric lifetime.

The reaction efficiencies for atmospheric species with smoke aerosol have not been measured. However, the reaction of OH, one of the most important oxidants in the atmosphere, with a graphite surface has been studied (Mulcahy and Young, 1975). Because the rate of the reaction was sufficiently fast to be diffusion limited in the experimental apparatus, only a lower limit for γ_{OH} of 5×10^{-2} was obtained. Although "wall effects" for other labile atmospheric species such as O_3, O, and HO_2 are well known because of the difficulties they pose in measurements of their homogeneous reaction rates, no γ values for reactions with atmospheric aerosol have been obtained.

It is not possible to make quantitative predictions of all the chemical composition changes of the troposphere following a nuclear war. However, it seems likely that the rate of oxidation of tropospheric species would be greatly decreased, particularly near the surface of the earth, for the period of time that the particulate matter resides in the atmosphere. Although oxidants in the atmosphere are usually looked upon as undesirable because of the damage they cause to plants and animals, oxidants serve an important function in cleansing the atmosphere of many anthropogenic and biogenic emissions. In fact, the lifetimes of nearly all compounds released to the atmosphere are determined by the rates of reaction with the hydroxyl radical. The source of OH radicals in the troposphere is photolysis of ozone, as discussed above. In addition to the reduced sunlight and loss of OH on particulate surfaces, OH concentrations would be reduced by combination with NO_2 to form nitric acid:

$$OH + NO_2 + M \rightarrow HNO_3 + M$$

In addition to the increased burden of toxic chemicals as the result nuclear war fires, one would expect large increases in the concentrations of many reduced compounds for two reasons: (1) the lifetimes of many compounds would be increased by large factors due to reduced concentrations of OH and other oxidants, and (2) biogenic emissions of some compounds might increase by large factors following a nuclear war. For example, compounds such as hydrogen sulfide and dimethyl sulfide are thought to have large biogenic emissions estimated at about 50 Tg S of each per year (Adams et al., 1981; Andreas and Raemdonck, 1983). However, their atmospheric concentrations are limited by short lifetimes of one or two days owing to reactions with the OH radical (Hatakeyama and Akimoto, 1983). It is difficult to predict the changes in biogenic emission rates that would follow a nuclear war. The stresses of the war on the biosphere, including a long period of darkness and freezing temperatures, would be expected to result in the death of many plants and animals, which in turn might lead to an increase in the rate of release of many reduced compounds. On the other hand, the low temperatures over land surfaces could decrease the rate of bacterial degradation of organic matter, and frozen freshwater systems could delay the escape of gaseous compounds to the atmosphere.

Because of the large heat capacity of the mixed layer of the ocean, the temperature of the ocean would be little changed. The principal effect of a nuclear war on biogenic emissions from the ocean would probably result from periods of low light intensity. Photosynthesis in the ocean takes place to a critical depth where the sunlight is attenuated to about 1 percent of its normal incident light flux. The darkness following a nuclear war would shift this critical depth much closer to the surface. As a result, one might expect the death of a significant fraction of the phytoplankton and zooplankton of the northern hemisphere ocean following a nuclear war (Milne and McKay, 1982).

Despite the large uncertainties, it is possible to place reasonable bounds on the concentrations of reduced sulfur compounds that would accumulate in the atmosphere. As a result of the rapid oxidation rate of dimethyl sulfide (DMS) in the normal atmosphere, the concentration of DMS in marine air is at least 2 orders of magnitude below the concentration that would be in equilibrium with seawater (Andreas and Raemdonck, 1983). As an upper bound, we may assume that the atmospheric concentration of DMS comes into equilibrium with surface water, resulting in an atmospheric mixing ratio of 21 ppbv. As a lower bound, we assume release of DMS at the present average sea-to-air flux (290 $\mu g\ S/m^2$ per day) for a period of 1 month and allow uniform mixing to an altitude of 10 km. This results in a mixing ratio of 0.8 ppbv. Considering that biogenic emissions of hydrogen sulfide are comparable in magnitude to DMS and that there would also be emissions from dimethyl disulfide and methyl mercaptan, for which emission factors are not well known, it appears likely that following nuclear war, the total concentration of reduced sulfur compounds in the troposphere would accumulate to a few parts-per-billion by volume. Although these are not toxic levels, at least for short-term exposure to humans, it is noteworthy that the threshold for smell in humans has been found to be in the ranges 0.9 to 8.5 ppbv for H_2S and 0.1 to 3.6 ppbv for $(CH_3)_2S$.

Toxic Chemical Releases

In addition to the emissions of carbon monoxide, nitrogen oxides, and organic compounds produced by the pyrolysis and partial combustion of wood, several million tons of noxious chemicals would be released to the atmosphere as a result of the pyrolysis and partial combustion of synthetic polymers such as rubber, plastics, and synthetic fibers located in urban areas, and chemicals in industrial storage. These chemical releases could have severe local consequences in and near the heavily populated urban areas. Occasional accidental releases of noxious chemicals have resulted in temporary evacuations of large areas. Contamination of the ground at very low levels (one part per million and below) by some particularly toxic chemicals has caused the permanent evacuation of some areas (e.g., Love Canal, New York, and Times Beach, Missouri). Recent attention has been drawn particularly to the polychlorinated biphenyls (PCBs), dioxins, and chlorine-substituted dibenzofurans. In the United States alone, more than 300,000 tons of PCBs are in use in electrical equipment and approximately 10,000 tons in storage (S. Miller, 1983). A large fraction of this toxic chemical could be released to the environment in a nuclear war. Apparently, dioxins and dibenzofurans may be produced in large quantities in the combustion of fuels containing chlorine, although this is currently a matter of considerable controversy (J.A. Miller, 1979; Bumb et al., 1980; Chemical and Engineering News, 1983).

Annual production in the United States of some important industrial chemicals is provided in Table 6.2. On the average, perhaps 5 to 10 percent of these amounts are in storage at any particular time. Pyrolysis and partial combustion of these and less abundant chemicals would result in the deposition of thousands of chemical species in the atmosphere and ultimately in the soil and water. The chlorine compounds would be expected to account for a large fraction of the more toxic, mutagenic, teratogenic, and carcinogenic compounds.

The problem of toxic chemicals released in a nuclear war is highly specific to locality and does not lend itself readily to general analysis. It seems likely, however, that portions of most of the urban areas affected would be seriously contaminated, at least in the smoky air during and immediately following burning. The possibility of serious local contamination of the ground and water for long times after the war cannot be ruled out.

Among the toxic materials released to the environment would be asbestos. The current world production of asbestos fibers amounts to about 4 million metric tons per year. More than 30 million tons (30 Tg) of asbestos has been accumulated in the United States alone. Accumulation by industrialized nations is in excess of 100 Tg. These fibers are bound in a wide variety of construction materials and other products. Much asbestos contained in the nonflammable materials would be released as the result of pulverization by the nuclear blast. Since asbestos fibers are nonflammable, they would also be released to the atmosphere upon combustion of materials such as floor tile and asphalt shingles.

It is difficult to estimate how much asbestos would be released to the atmosphere as the result of a nuclear war. However, when mixed uniformly throughout the lower 9 km of the atmosphere and over half of the northern hemisphere, the atmospheric concentration of asbestos is calculated to be about 0.3 fibers per cubic centimeter for each teragram of asbestos released. This calculation uses the conversion factor used in epidemiological studies in which it is assumed that 1 fiber would be detected by phase contrast light microscopy for every 30 x 10^{-12} g of suspended asbestos. An optical fiber is defined as any particle longer than 5 μm, having a length-to-diameter ratio of at least 3-to-1 and a maximum diameter of 5 μm. Of course, the actual number of fibers is much larger, owing to the preponderance of smaller fibers not counted. The present Occupational Safety and Health Administration (OSHA) standard for exposure to asbestos is a time-weighted average of 2.0 fibers per cubic centimeter over an 8-h period, and OSHA announced a decision to lower it to 0.5 fiber per cubic centimeter in November 1983. A recent NRC study (NRC, 1984b) estimated the average nonoccupational exposure in the United States to asbestos to be 0.0004 fibers per cubic centimeter. Five teragrams (less than 5 percent of the world accumulation) of asbestos released to the atmosphere would increase the general population exposure to asbestos by a factor of about 4000 for the period of time that the particles are suspended and uniformly distributed. Of course, the fibers would be subject to resuspension and would be concentrated in the boundary layer of the atmosphere.

TABLE 6.2 U.S. Production of Some Major Chemicals in 1982

	Millions of Tons
Sulfuric acid	29.4
Ammonia	14.1
Ethylene	11.2
Chlorine	8.3
Phosphoric acid	7.8
Toluene	6.9
Nitric acid	6.9
Propylene	5.6
Ethylene dichloride	4.5
Xylenes	3.8
Benzene	3.6
Methanol	3.3
Ethylbenzene	3.0
Vinyl chloride	3.0
Styrene	2.7
Hydrochloric acid	2.4
Terephthalic acid	2.3
Ethylene oxide	2.2
Ethylene glycol	1.8
Acetic acid	1.2
Cumene	1.2
Phenol	0.96
Acrylonitrile	0.92
Vinyl acetate	0.85
Butadiene	0.83
Acetone	0.80
Formaldehyde	0.76
Propylene oxide	0.67
Isopropanol	0.59
Cyclohexane	0.58
Adipic acid	0.54
Acetic anhydride	0.48
Ethanol	0.46

REFERENCES

Adams, D.F., S.O. Farwell, E. Robinson, M.R. Pack, and W.L. Bamesberger (1981) Biogenic sulfur source strengths. Environ. Sci. Technol. 15:493-498.

Andreas, M.O., and H. Raemdonck (1983) Dimethyl sulfide in the surface ocean and the marine atmosphere: A global view. Science 221:744-747.

Ambio (1982) Nuclear war: The aftermath. 11(2/3):75-176.

Angell, J.K., and J. Korshover (1973) Quasi-biennial and long-term fluctuation in total ozones. Mon. Weather Rev. 101:426, 104:63.

Birks, J.W., and J. Staehelin (1984) Changes in tropospheric composition and chemistry resulting from a nuclear war. Draft manuscript.

Bumb, R.R., W.B. Crummett, S.S. Cutie, J.R. Gledhill, R.H. Hummel, R.O. Kagel, L.L. Lamparski, E.V. Luoma, D.L. Miller, T.J. Nestrick, L.A. Shadoff, R.H. Stehl, and J.S. Woods (1980) Trace chemistries of fire: A source of chlorinated dioxins. Science 210:385-389.

Chang, J.S., and D.J. Wuebbles (1982) The Consequences of Nuclear War on the Global Environment. Hearing before the Subcommittee on Investigations and Oversight of the Committee on Science and Technology, U.S. House of Representatives, Sept. 15.

Chang, J.S., W.H. Duewer, and D.J. Wuebbles (1979) The atmospheric nuclear test of the 1950's and 1960's: A possible test of ozone depletion theories. J. Geophys. Res. 84:1755.

Chemical and Engineering News (1983) Special Issue on Dioxin. June 6.

Chi, C.T., et al. (1979) Source Assessment: Prescribed Burning, State of the Art. EPA Report EPA-600/2-79-019h. Research Triangle Park, N.C.: U.S. Environmental Protection Agency.

Clements, H.B., and C.K. McMahon (1980) Nitrogen oxides from burning forest fuels examined by thermogravimetry and evolved gas analysis. Thermochim. Acta 35:133.

Crutzen, P.J. (1971) Ozone production rate in an oxygen-hydrogen oxide atmosphere. J. Geophys. Res. 76:7311.

Crutzen, P.J., and J.W. Birks (1982) The atmosphere after a nuclear war: Twilight at noon. Ambio 11:114-125.

DeAngelis, D.G, D.S. Ruffin, and R.B. Reznik (1980) Preliminary Characterization of Emissions from Wood-fired Residential Combustion Equipment. EPA Report 600/7-80-040. Research Triangle Park, N.C.: U.S. Environmental Protection Agency.

Foley, H.M., and M.A. Ruderman (1973) Stratospheric NO production from past nuclear explosions. J. Geophys. Res. 78:4441.

Gilmore, F.R. (1975) The production of nitrogen oxides by low-altitude nuclear explosions. J. Geophys. Res. 80:4553.

Greenberg, J.P., P.R. Zimmerman, L. Heidt, and W. Pollack (1984) Hydrocarbon emissions from biomass burning in Brazil. J. Geophys. Res. 89:1350-1354.

Hatakeyama, S., and H. Akimoto (1983) Reactions of OH radicals with methanethiol, dimethyl sulfide, and dimethyl disulfide in air. J. Phys. Chem. 87:2387-2395.

Johnston, H.S. (1971) Reduction of stratospheric ozone by nitrogen oxide catalysts from supersonic transport exhaust. Science 173:517.

Johnston, H.S., G. Whitten, and J.W. Birks (1973) Effects of nuclear explosions on stratospheric nitric oxide and ozone. J. Geophys. Res. 78:6107.

Logan, J.A., M.J. Prather, S.C. Wofsy, and M.B. McElroy (1978) Atmospheric chemistry: Response to human influence. Phil. Trans. Roy. Soc. London 290:187.

Luther, F.M. (1983) Nuclear war: Short-term chemical and radioactive effects of stratospheric injections. Paper presented at the International Seminar on Nuclear War, 3rd Session: The Technical Basis for Peace. Ettore Majorana Centre for Scientific Culture, Erice, Sicily, August 19-24, 1983.

Mahlman, J.D., and W.J. Moxim (1978) Tracer simulations using a global circulation model: Results from a mid-latitude instantaneous source experiment. J. Atmos. Sci. 35:1340-1374.

McMahon, C.K. (1983) Characterization of forest fuels, fires, and emissions. Paper presented at the 76th Annual Meeting, Air Pollution Control Association, Atlanta, Ga.

Miller, J.A. (1979) Chemists disagree on dioxin sources. New Sci. (Oct. 4):25.

Miller, S. (1983) The PCB imbroglio. Environ. Sci. Technol. 17:11A-14A.

Milne, D.H., and C.P. McKay (1982) Response of marine plankton communities to a global atmospheric darkening. Geol. Soc. Am. Spec. Pap. 190:297-303.

Mulcahy, M.F.R., and B.C. Young (1975) Heterogeneous reactions of OH radical. Int. J. Chem. Kinet. 7(suppl.):595-609.

National Research Council (1973) Climatic Effects of Supersonic Flight. Washington, D.C.: National Academy of Sciences.

National Research Council (1975) Long-Term Worldwide Effects of Multiple Nuclear Weapon Detonations. Washington, D.C.: National Academy of Sciences.

National Research Council (1982) Causes and Effects of Stratospheric Ozone Reduction: An Update. Washington, D.C.: National Academy Press.

National Research Council (1984a) Causes and Effects of Changes in Stratospheric Ozone: Update 1983. Washington, D.C.: National Academy Press.

National Research Council (1984b) Asbestiform Fibers: Nonoccupational Health Risks. Washington, D.C.: National Academy Press.

Ogren, J.A. (1982) Deposition of particulate elemental carbon from the atmosphere. Pages 370-391 in Particulate Carbon: Atmospheric Life Cycle, edited by G.T. Wolff and R.L. Klimisch. New York: Plenum Press.

Ryan, P.W., and C.K. McMahon (1976) Some chemical and physical characteristics of emissions from forest fires. Paper 76-2.3 presented at the 69th Annual Meeting, Air Pollution Control Association, Portland, Ore.

Sandberg, D.V., S.G. Pickford, and E.F. Darley (1975) Emissions from slash burning and the influence of flame retardant chemicals. J. Air Pollut. Control Assoc. 25:278-281.

Sedlacek, W.A., E.J. Mroz, A.L. Lazrus, and B.W. Gandrud (1983) Various Nitric Acid Concentrations in the Lower Stratosphere: 1971-1982. Report LA-UR-83-3138. Los Alamos, N.Mex.: Los Alamos National Laboratory.

Tangren, C.D., C.K. McMahon, and P.W. Ryan (1976) Contents and effects of forest fire smoke. Southern Forestry Smoke Management Guidebook. U.S. For. Serv. Gen. Tech. Rep. SE-10.

Turco, R.P., O.B. Toon, T.P. Ackerman, J.B. Pollack, and C. Sagan (1983) Global Atmospheric Consequences of Nuclear War. Interim Report. Marina del Rey, Calif.: R&D Associates. 144 pp.

Ward, E.E., et al. (1982) Measurement of smoke from two prescribed fires in the Pacific Northwest. Paper 82-8,4 presented at the 75th Annual Meeting, Air Pollution Control Association, New Orleans, La.

7 Atmospheric Effects and Interactions

OVERVIEW

The dispersion, evolution, and effects of dust and smoke injected into the atmosphere from a major nuclear conflict involve a large set of interacting processes whose complexity precludes detailed quantitative prediction at the present time. The available tools include a variety of models, of which the most advanced are the general circulation models (GCMs) developed for application to studies of weather prediction and climate dynamics. In these models, pressure, temperature, wind, moisture, and cloudiness fields are represented with a horizontal resolution of a few hundred kilometers and at a number of tropospheric and stratospheric levels (see, for example, Gates and Schlesinger, 1977; Mahlman and Moxim, 1978; Washington, 1982). Smaller scale processes such as microscale and mesoscale turbulence, convection, gravity waves, local topography, and land-sea circulations can only be treated parametrically. Nevertheless, several of these models provide realistic simulations of the present climate.

For applications to the problem of atmospheric effects of dust and smoke from nuclear war, however, GCMs are deficient in several respects. Transport of trace gases and diurnal variations have been simulated in some GCM studies (Levy et al., 1980; Cess et al., 1984; MacCracken and Walton, 1984). However, no existing GCM simulates the full physics of a radiatively active trace material where net heating effects drive the circulation while the distribution of material is itself continuously varying in response to the flow and to complex flow-dependent removal processes. Formulations of boundary layer processes in these models are necessarily somewhat crude because of the low spatial resolution. Some recent model calculations have included particulate transport and diurnally varying absorption of solar radiation by the particulates, but these calculations have thus far had very limited vertical resolution (Cess et al., 1984; MacCracken and Walton, 1984). Perhaps most serious for the nuclear war particulate problem, the cloud microphysical processes that are primarily responsible for the removal of particulates from the atmosphere cannot now be included in detail in these models.

Other more specialized models can be applied to aspects of the problem, for example: cumulus-scale and mesoscale circulation models, some with crude treatments of cloud microphysics, could be used to investigate specific processes that occur at scales smaller than that of the GCM grids. One-dimensional (vertical) radiative-convective models coupled to particle microphysical models have been used for detailed investigations of these critical processes, and, because of their computational efficiency, such models are extremely useful for sensitivity studies. Two-dimensional circulation models, though far less realistic than GCMs, can simulate the zonally symmetric components of the flow and the corresponding transport and radiative heating effects of nuclear particulates. Because they are relatively convenient computationally, they can be used for sensitivity studies, and therefore provide a valuable complement to GCMs.

Energy balance climate models (EBCMs) make up another class of relatively simple model that can be used to investigate radiative perturbations of surface energy balance and surface temperature (e.g., Sellers, 1973; Robock, 1983). Most such models deal only with the energy balance at the surface, and horizontal heat transport is modeled as a diffusive process with diffusion coefficients chosen to provide reasonable simulations of the present climate. Consequently, results from such models must be interpreted judiciously. The advantage of EBCMs is that, because of their computational efficiency and modeling of horizontal variations, they can be used to provide an indicator of the feedback effects of such relatively persistent climate factors as snow and ice albedo, sea ice cover, and sea surface temperature.

Some of the principal results that are now available from one-, two-, and three-dimensional models are displayed in Tables 7.3 and 7.4.

Of necessity, the results of simulations using models constitute the core of our knowledge of the likely atmospheric effects of smoke and dust from a nuclear war. In discussing these results, it is convenient to divide the problem into several subdivisions: early spread and evolution of the particulate clouds, direct optical effects, thermal effects as calculated by one-dimensional (vertical) models, thermal and circulation effects calculated by multidimensional models, and modification of circulation, cloudiness, and precipitation fields by the radiation perturbations induced by these particulate clouds. Several of these are rapidly evolving areas of research, and it should be clear that parts of this chapter may be superseded by new developments in the near future.

In the absence of observational analogs of the atmosphere as perturbed by nuclear war, observations of related, though inevitably very different atmospheric situations must be used. Several such partial analogs are discussed near the end of this chapter. The global-scale atmospheric perturbations associated with major volcanic eruptions and with plausible meteor impact events and their relationship to nuclear war scenarios are considered in the following chapter.

EARLY SPREAD AND EVOLUTION OF PARTICULATE CLOUDS

The area initially covered by the smoke plumes would depend on the number of fires, the cross-wind width of each fire, the average wind speed, the directional variability of the wind near the level of plume stabilization, the duration of the fires, and the overlap among fire zones. If urban fire plumes extended into the middle troposphere, they would be transported by winds whose average speeds are of order 20 m/s, so that fires of several hours duration would produce plumes several hundred kilometers in length. For this reason alone, it is reasonable that substantial fractions of Eurasia, North America, and the North Atlantic, would be covered initially by smoke plumes. Crutzen et al. (1984) have estimated that the initial area covered by smoke plumes would be between 1×10^7 km^2 and 2×10^7 km^2 for a scenario similar to the *Ambio* scenario (*Ambio*, 1982). For the committee's 6500-Mt scenario with about 1000 urban mass fires, an initial coverage area (immediately following the phase of rapid burning and plume rise) of about 10^7 km^2 seems to be reasonable.

In a statically stable atmosphere subject to solar heating, local wind systems would develop in response to the differential heating associated with nonuniformities in smoke distribution, and these winds would tend to smooth out both the thermal perturbations and the smoke nonuniformities. Such forced circulation systems were found to be effective smoothing agents in a cumulus-scale circulation model with an initially nonuniform distribution of carbon black (Chen and Orville, 1977). The committee is not aware of similar numerical experiments at larger scales, but there is good reason to believe that such wind systems would be effective at scales out to several hundred kilometers. This is the typical scale of the Rossby radius of deformation.* At this and larger scales, the effect of earth's rotation becomes important and would impose a structure that could partially restrict such thermally forced lateral spreading of the smoke. Nevertheless, many of the smoke-free holes originating over the North American and Eurasian continents between 30°N and 70°N latitude would be filled within the first 2 days.

After about 3 days, under typical meteorological conditions, the major gap over the Atlantic in the 30°N to 60°N latitude belt would be largely filled and very likely would have drifted over Western Europe. Portions of the mid-latitude Pacific would also be covered. The speed of this further spreading would depend somewhat on season, being greater in winter and smaller in summer. Figure 7.1 shows a specific winter season realization of the smoke and dust distribution after 3 days, based on winds derived from the Oregon State GCM (Gates and Schlesinger, 1977), and a nuclear war smoke injection scenario somewhat similar to the baseline case (MacCracken, 1983). The initial injections for this case were 207 Tg soot and 118 Tg dust. The feedback between the radiatively induced perturbation to circulation and particulate transport was not included; it was, however, included in a more recent

---

*The Rossby radius of deformation for mid-latitude disturbances driven by heating in the mid-troposphere is $(N/f)H$, where $N \sim 10^{-2}$ s^{-1} is the frequency of buoyancy oscillations, $f \sim 10^{-4}$ s^{-1} is the Coriolis frequency, and $H \sim 0.7 \times 10^4$ m is the scale height (e.g., Holton, 1979). Hence the Rossby radius is about 700 km.

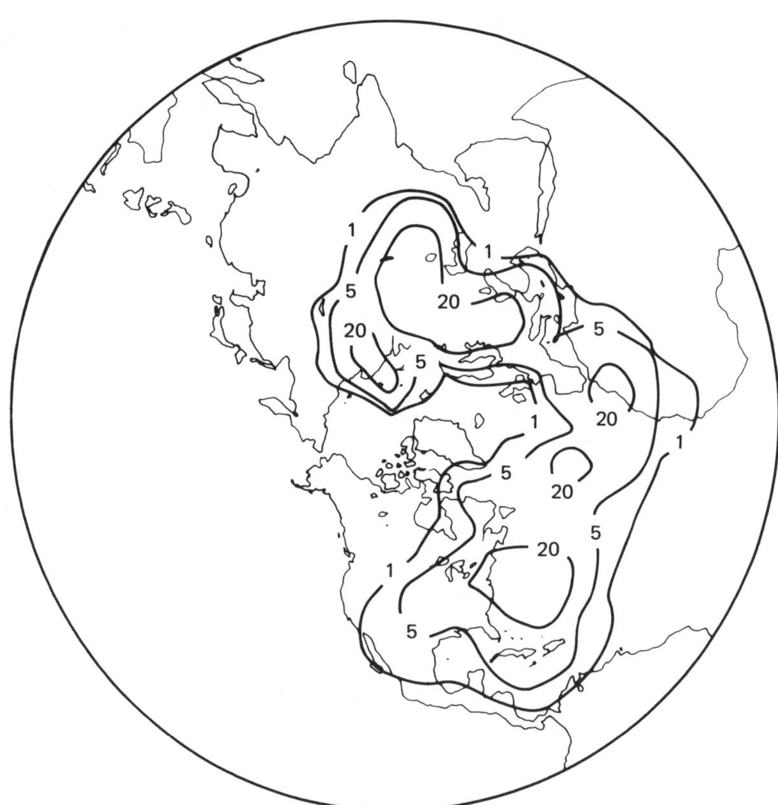

FIGURE 7.1 Hemispheric distribution of smoke-induced optical depth 3 days after a hypothetical nuclear exchange. (From MacCracken, 1983.)

calculation using the Oregon State GCM, which produced quite similar results (MacCracken and Walton, 1984). The spreading of smoke is probably underestimated in the calculation shown in Figure 7.1 because vertical wind shear has been neglected and the thermally forced smoothing and spreading of the smoke have not been taken into account. Nevertheless, smoke and dust cover much of the northern mid-latitude region. According to this calculation, there are large patches (of order 10^6 km^2 in area) in which optical depth exceeds 20 at 3 days after the start of fires, but approximately 20 percent of the area of the hemisphere (about 40 x 10^6 km^2) is already covered by smoke and dust with optical depth of 5 or more.

The initial area covered by stratospheric dust, corresponding to the area occupied initially by stabilized nuclear clouds, would be much smaller, about 0.4 x 10^6 km^2 for the baseline case. Although dust absorbs solar radiation far less efficiently than smoke, the heating per unit mass of air would still be significant at the lower densities of the stratosphere. Thus these clouds would also tend to spread laterally in response to their self-induced thermal circulation.

Calculations of stratospheric dust cloud dispersion for a nuclear war scenario involving counterforce strikes show distributions qualitatively similar to that in Figure 7.1 when climatological mean midwinter winds are used (B. Yoon, private communication, 1983).

Dispersion would be faster with actual time-dependent winter winds, but during summer, spring, and autumn, zonal winds in the extratropical lower stratosphere are weaker, and dispersion would be correspondingly slower. Material injected above 18 km in midsummer would drift westward (e.g., Holton, 1975).

As discussed on pages 77 to 80, nuclear smoke clouds would be subject to early rainout and coagulation of particles during the initial plume rise phase, but the effectiveness of these processes would rapidly decrease after the clouds have stabilized and begun to spread out in horizontally stratified plumes. Crutzen et al. (1984), using a simplified model, found less than a factor of 2 increase of particle mode radius during the 30 days following the initial rapid rise phase of the fire plumes. Coagulation in slowly dispersing smoke clouds was also evaluated by Turco et al. (1983b). In a case intended to maximize the Brownian coagulation rate, they assumed initial plume coverage equal to that of the stabilized nuclear clouds (about 10^6 km^2 for their baseline case); they also assumed slow horizontal diffusive growth such that coverage increased linearly with time, reaching 20 x 10^6 km^2 only after 20 days. For the reasons cited above, this spreading rate is unrealistically slow, but even with these extreme assumptions, average smoke particle radius was found to increase by only about 65 percent after 1 week.

For spherical particles whose initial radii are $\lesssim 0.4$ μm having an imaginary refractive index (the absorption component of refractive index) of $\lesssim 0.1$, such size increases cause a decrease in the absorption coefficient per unit mass of less than a factor of 2 (Bergstrom, 1973; Lee, 1983). For smaller or more weakly absorbing particles and for infrared radiation, the effect of such a size change is smaller. As will be shown below, early temperature changes near the surface are not very sensitive to variations of a factor of 2 or less from the absorption coefficient per unit mass of the baseline smoke injections. This is because the baseline injection initially contains more than enough smoke to absorb almost all sunlight in the areas affected by the smoke cloud. The duration of direct thermal effects of the particulates is more sensitive to the absorption coefficient, however. In addition, factor of 2 reductions below the baseline in several quantities (e.g., initial injected mass and absorption coefficient) would affect even the early temperature changes. Thus coagulation and early rainout are very important and complex issues requiring additional research.

Longer term chemical and physical modification, or "aging," of aerosols in the atmosphere is another area on which additional basic information is needed. Because elemental carbon is hygrophobic and unreactive in the atmospheric temperature range, this may be a slow process for soot, depending on coalescence with preexisting hygroscopic particles. When coalescence occurs, the resulting particles behave as hygroscopic particles and can grow further by adsorption of water (Ogren, 1982; Ogren and Charlson, 1983). Because of the internally mixed elemental carbon, these composite particles would still be efficient absorbers of sunlight (Ackerman and Toon, 1981), but an increase in composite particle size due to aging could have an important effect on the ratio of absorption efficiencies at visible and infrared wavelengths. Since this ratio is an important factor controlling the influence of particulates on net radiation, the aging issue requires careful additional scrutiny.

DIRECT OPTICAL EFFECTS

Figure 7.2 displays the transmission of visible sunlight, including diffuse as well as direct radiation, as a function of smoke and dust opacities for particulates having the size and refractive index properties specified in the baseline case. Dust and smoke properties for the injections of the baseline case have been described and presented in Chapters 4 and 5 (readers are referred particularly to pages 27 to 32 and Table 5.7). For convenience, the baseline injection parameters are summarized in Table 7.1.*,† For these optical properties, light levels decrease very rapidly for smoke optical depths greater than one. When these light level reductions are combined with the extinction optical depths calculated by MacCracken (1983), and illustrated in Figure 7.1, the result is that light levels for much of the continental area north of 30°N would be reduced below the limit of photosynthesis during the first week, and widespread dense patches of smoke would make seeing impossible for several days after the nuclear exchange. For the NRC baseline case, with smoke and dust assumed to be instantaneously dispersed to a uniform distribution over the 30°N to 70°N latitude belt, average light levels for the belt would be below those for a very cloudy day (about 10 percent of the normal clear sky illumination) for about 2 weeks after the exchange. This can be seen by comparing the total downward solar flux versus time for this case (Figure 7.3) with the transmission levels shown in Figure 7.2.

The values shown in Figure 7.3 were calculated using the one-dimensional model of Turco et al. (1983a,b). As explained by these authors, this model combines a detailed radiative transfer model with a detailed particle microphysics model (Pollack et al., 1976; Toon et al., 1979; Turco et al., 1979; Ackerman and Toon, 1981; Pollack et al., 1983).

There is an approximately exponential dependence of the total downward solar flux on smoke opacity when full allowance is made for multiple scattering, as shown in Figure 7.2. This is largely because of the high absorptivity of the smoke. As a consequence, a saturation effect occurs such that most of the solar flux is removed by a smoke optical depth as small as 2; further increases in smoke optical depth

*Readers unfamiliar with radiative transfer theory may wish to consult Liou (1980), which describes the theory and computational approaches in detail.

†The abbreviation NRC is used to denote the committee's baseline and excursions; LLNL denotes Lawrence Livermore Laboratory (e.g., MacCracken, 1983), and TTAPS denotes Turco et al. (1983a,b).

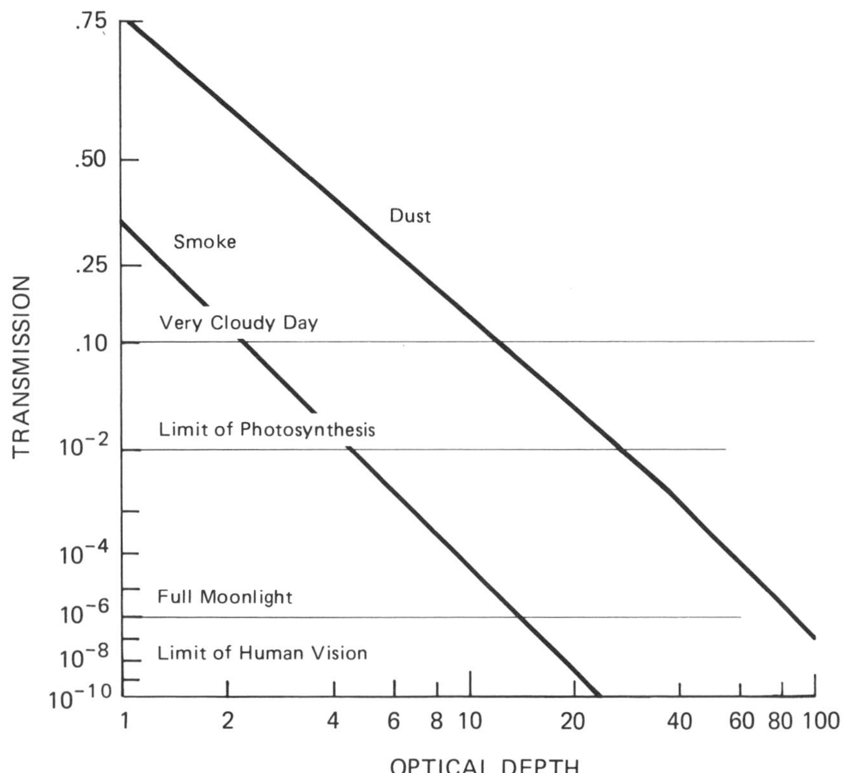

FIGURE 7.2 Fraction of incident solar radiation reaching the surface as a function of extinction optical depth for smoke and dust particulates with optical properties as in the NRC baseline case (Table 7.1). Solar zenith angle of 60° is assumed. Diurnally averaged illumination depletions would be somewhat smaller at latitudes and seasons with smaller minimum zenith angles. These calculations use the radiative transfer algorithm of Pollack et al. (1976, 1983), in which full account is taken of multiply scattered radiation (cf. Pollack et al., 1983, and references therein for a fuller description). Note that the vertical scale is logarithmic.

have relatively little additional effect on solar flux received at the surface. This saturation effect carries through to the temperature changes computed by one-dimensional radiative-convective models (Turco et al., 1983a) and energy balance climate models (Robock, 1984), since the degree of cooling at the surface predicted by these models is not very sensitive to variations in illumination at very low light levels, and these models do not allow for gaps and nonuniformity in the smoke.

Because of the high absorptivity, smoke clouds produce much larger depletions of solar radiation than water clouds or dust clouds of comparable extinction optical depth. However, even for a relatively moderate depletion in surface illumination comparable to that produced by dense water clouds, smoke clouds would have a larger effect on the surface thermal balance than water clouds. This is because water clouds have a high ratio of infrared to visible opacity so that increased

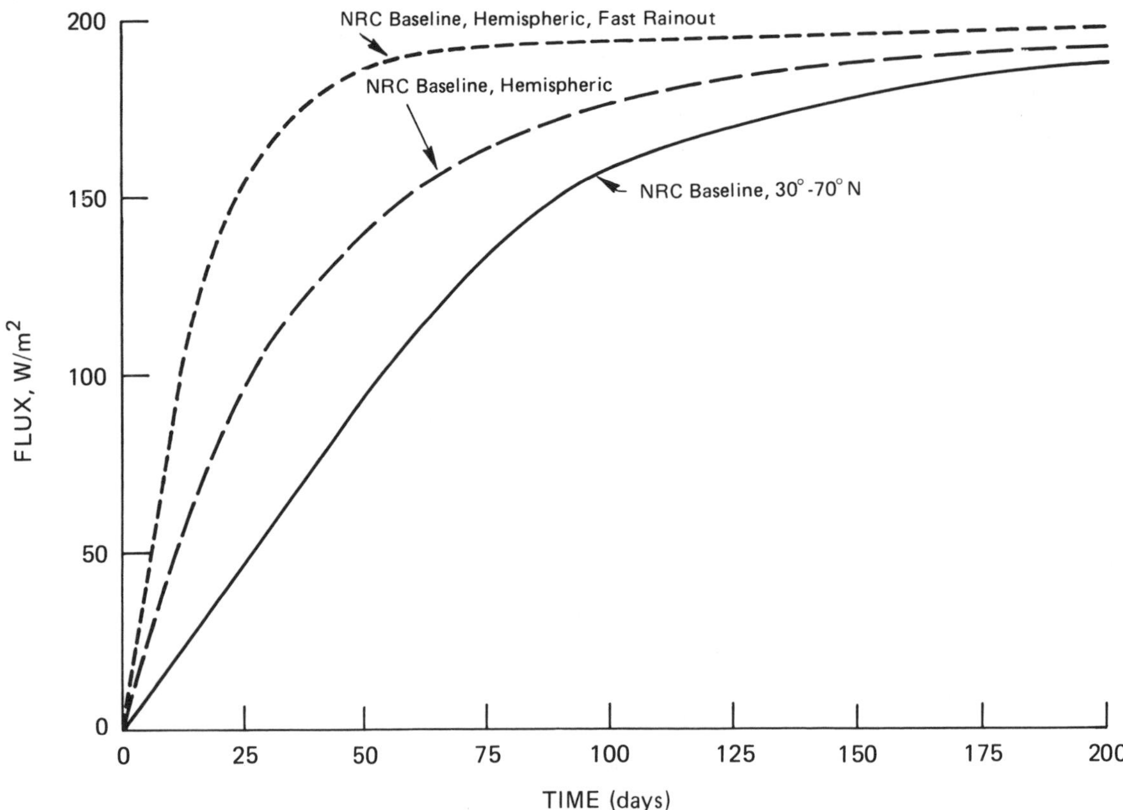

FIGURE 7.3 Time evolution of the total downward solar flux at the surface for the NRC baseline (30°N to 70°N), the NRC baseline injections spread over 0° to 90°N, and the NRC fast-rainout variant with injections spread over 0° to 90°N.

downward flux of infrared radiation can equal, or even exceed (on a 24-h basis), the depletion of solar flux. Such compensation between solar radiation depletion and infrared radiation enhancement would not occur for smoke clouds because of their low ratio of infrared to visible opacity, except in regions where the normal daily total of solar radiation is already very low, such as is the case very close to the polar twilight boundary during winter, or would be the case at very early times following a nuclear exchange in dense patches in which the optical depth reaches values of 20 or more.

The corresponding saturation regime is not reached for dust until the optical depth of dust alone reaches a value of about 12 (see Figure 7.2). For this reason, among others, the thermal effect of dust is far more sensitive than that of smoke to the nuclear war scenario. Smoke opacity is initially well within the saturation regime for the baseline smoke emission given in Table 7.1--180 Tg spread over the 30°N-70°N latitude belt--and approaches the margin of the saturation regime only as this value is decreased by about a factor of 4 to ~40 to 50 Tg. For smoke injections below this level, saturation no longer applies, and the light reduction and temperature effects would decrease rapidly with

TABLE 7.1 Properties of Injected Aerosols, NRC Baseline Case

	Dust (see pages 27 to 32)	Smoke (see Table 5.7)
Total injected mass (Tg)	15	180
Median particle radius r_m (μm)[a]	0.25	0.10
Log normal dispersion γ[a]	2.0	2.0
Refractive index (real part, 0.5 μm)	1.5	1.55
Refractive index (imaginary part, 0.5 μm)	0.001	0.10
Extinction coefficient at 0.5 μm (m^2/g)	2.8	5.5
Absorption coefficient at 0.5 μm (m^2/g)	0.1	2.0
Infrared optical properties	Wavelength-dependent basaltic glass (cf. Pollack et al., 1973)	Absorption only, cross section 0.5 m^2/g
Vertical distribution of injection	37% stratosphere 63% troposphere (see Table 4.1)	Uniform mass per unit volume between 0 and 9 km (see pages 73 to 76 and 83)
Horizontal distribution of injection	Uniform in the latitude belt 30°N-70°N; none outside	Uniform in the latitude belt 30°N-70°N; none outside

[a]Parameters of the log normal size distribution; see page 62.

decreasing injected mass. On the other hand, dust opacity approaches saturation only for rather extreme excursions that involve large numbers of surface bursts. For the dust optical properties and quantities of the baseline case, the extinction cross section of dust at 0.5 μm is 2.8 m^2/g, and the corresponding extinction optical depths [lower limit (best estimate) upper limit for submicron dust in both troposphere and stratosphere] are [0.6 (0.9) 1.5] for dust uniformly spread around the 30°N to 70°N latitude belt. The extinction optical depths in the same belt with the added opacity due to the 8500-Mt dust excursion (see page 32) are [1.3 (2.1) 3.3]. The values for the 8500-Mt excursion, though well below the saturation threshold for dust, may nevertheless be climatologically significant since most of this dust is in the stratosphere and has a long residence time. In the baseline case, only about 40 percent of the dust is initially injected into the stratosphere, but the remainder may have an anomalously long residence time in the upper troposphere if precipitation is suppressed because of the smoke (see below).

Crutzen et al. (1984) have also estimated transmission versus time for the smoke cloud. They consider models with rainout removal times of 15 days and 30 days. For the 15-day rainout time, calculated solar illumination reductions to the 10 percent level persist for 10 to 14 days (the exact value depending on the assumed extent of forest fires), by which time a uniform cloud has dispersed to cover 60 percent of the northern hemisphere, whereas for the 30-day rainout time the reduction to 10 percent persists for about 14 to 24 days. According to their estimates, about one-half of the northern hemisphere will have been covered by the smoke cloud in 10 days, and about two-thirds of the hemisphere in 20 days. These reductions correspond quite well to the NRC baseline case despite differences in the scenarios and in the treatment of cloud dispersion and evolution.

THERMAL EFFECTS IN ONE-DIMENSIONAL MODELS

General circulation models can provide the most detailed and reliable assessments of temperature changes associated with nuclear war; however, because of their complexity and computational requirements, they are not suitable for sensitivity studies in which parameters such as input scenarios and particulate removal rates are varied over wide ranges. Turco et al. (1983a,b) have carried out such sensitivity studies using the TTAPS one-dimensional model. In order to relate the results of the TTAPS studies to the current baseline and to the results of multidimensional modeling studies using the NRC baseline, the TTAPS model has been applied to the NRC baseline case and to two variations: a fast-rainout removal case, and a case in which the baseline smoke injection is uniformly distributed over the entire northern hemisphere.

The TTAPS one-dimensional model (Turco et al., 1983b) calculates the microphysical evolution of particulates subject to coagulation, agglomeration, sedimentation, vertical eddy diffusion, surface deposition, and removal by parameterized rainout processes (see Turco et al., 1983b, and references therein--particularly Turco et al., 1979, 1981; Toon et al., 1979; Hamill et al., 1982--for details). In the NRC baseline case the smoke and dust clouds have been assumed to be uniformly distributed around the 30°N to 70°N latitude belt. The microphysical implications of this simplifying assumption have been discussed in previous sections of this chapter and in Chapter 5. Results are most sensitive to the particle process, which is parameterized as a linear loss mechanism with a height-dependent exponential lifetime. Since the particle lifetime increases rapidly with altitude, there is a strong interaction between the altitude of initial smoke plume injection and the assumed vertical profile of rainout rate. As described on pages 73 to 76 the committee has assumed that the smoke is distributed uniformly with altitude over the 0 to 9 km range, partly for simplicity in the absence of better information, and partly because it is the committee's judgment that the intensity of urban fires would tend to drive the plumes into the upper troposphere. If vertical mixing in the plumes is very rapid, it would tend to produce a uniform smoke mixing ratio rather than uniform smoke concentration. However, as will be seen below, the tendency to develop a uniform mixing ratio would probably decrease rapidly with time and would be strongly opposed by the increase in the rainout rate near the ground.

The rainout removal rate profile assumed for the NRC baseline case is given in Table 7.2, where it is compared with the profile used in the TTAPS study. The TTAPS group chose baseline values designed to represent the rainout characteristics of the unperturbed atmosphere; for the NRC baseline case, these values have been modified so that faster rainout occurs in the lower troposphere (0 to 5 km) and no rainout at all occurs above 5 km. These changes have been made in order to simulate possible effects of changes in static stability and cloudiness expected in the perturbed atmosphere (see pp. 156 to 158 below), and they are of course highly uncertain. Even in the absence of rainout, however, eddy diffusion acts in the model as an effective mechanism for removing particulates from the upper troposphere. Following Massie and Hunten (1981), the vertical eddy diffusion coefficient value 10 m^2/s has been assumed for the NRC baseline case, as it was for the TTAPS calculations. This value gives a characteristic lifetime against dry deposition for particulates in the upper troposphere of about 40 days.

Because the rainout time and its interaction with the initial vertical smoke distribution are so critical to the evaluation of climatic effects, a fast-rainout excursion has been considered, with rainout times given in the last column of Table 7.2. These high values of rainout rate are believed to provide a reasonable case bounding the smoke lifetime on the low side for the NRC baseline smoke injection. In this case, the smoke has been assumed to be dispersed over the entire hemisphere rather than over the 30°N to 70°N latitude band. The initial opacity for this case is nearly equivalent to that for smoke and dust spread over the 30°N to 70°N latitude band with half of the initial smoke and dust injections of the NRC baseline case. The TTAPS "slow-rainout" case, with an effective removal rate about one-third as fast as their baseline case, represents a plausible bound to the smoke lifetime on the high side. This case is also compared with the NRC baseline.

Figure 7.4 shows vertical profiles of the contributions to optical depth from smoke and dust in each 2-km layer for (a) the NRC baseline case, and (b) the fast-rainout excursion. For comparison, the TTAPS baseline is also shown (Figure 7.4c). In the TTAPS baseline case, the relatively rapid rainout removal assumed for the upper troposphere causes the center of mass of smoke to lower over time, while in both the NRC baseline case and the fast-rainout excursion, the rapid downward

TABLE 7.2 Smoke Removal Rates (s^{-1})

Altitude (km)	TTAPS (1983, 1984)	NRC Baseline	NRC "Fast-Rainout" Case
0-1	1.0×10^{-6}	4.0×10^{-6}	4.0×10^{-6}
1-3	8.6×10^{-7}	2.7×10^{-6}	4.0×10^{-6}
3-5	7.1×10^{-7}	1.3×10^{-6}	2.0×10^{-6}
5-7	5.7×10^{-7}	0	1.0×10^{-6}
7-9	4.3×10^{-7}	0	1.0×10^{-6}
9-11	2.9×10^{-7}	0	1.0×10^{-6}
11-13	1.4×10^{-7}	0	0

NOTE: Values represent [(1/m)(dm/dt)], where m is the mass of particulates per unit volume. Zeroes indicate removal controlled by eddy diffusion with $K = 10\ m^2/s$. Smoke from above 5 km diffuses downward to 5 km, where it is removed by the rates shown. Effective removal rates for the atmosphere above 5 km can be estimated from the data shown in Figure 7.4a, and amount to about $3 \times 10^{-7}\ s^{-1}$ for the layer between 5 and 8 km for the first 30 days following smoke injection.

increase in rainout rate quickly removes smoke in the lower troposphere, leaving an elevated smoke cloud. Note that the low initial values of opacity in the TTAPS baseline and in the NRC fast-rainout varient are due to the assumption of initial dispersion over the hemisphere for these cases. The NRC baseline also differs from the TTAPS baseline in that the latter has smoke in the lower stratosphere and a larger mass of stratospheric dust. These features appear as bulges near 15 and 25 km in the initial profiles of Figure 7.4c. The differences in shape below 10 km between the TTAPS and NRC cases are due to differences in assumed initial injections and removal rate profiles. Figure 7.5 shows the time evolution of the optical depth for the NRC baseline, the fast-rainout excursion, and the TTAPS baseline. The total optical depth with contributions from dust as well as smoke is shown. These results reflect the removal rates shown in Table 7.2.

The radiative-convective component of the TTAPS model has been described elsewhere, and readers are referred to these sources for details (Pollack et al., 1976; Ackerman and Toon, 1981; Toon and Ackerman, 1981; Pollack et al., 1983). The model takes into account the size distributions and optical properties of smoke and dust, and explicitly calculates solar and infrared radiation including the effects of multiple scattering, absorption and emission by atmospheric gases, and the radiative effects of a prescribed representative distribution of cloudiness. Heat capacity of the underlying surface is neglected for land areas, but is included when the model is applied to ocean areas as it was in one of the cases considered by Turco et al. (1983a). The

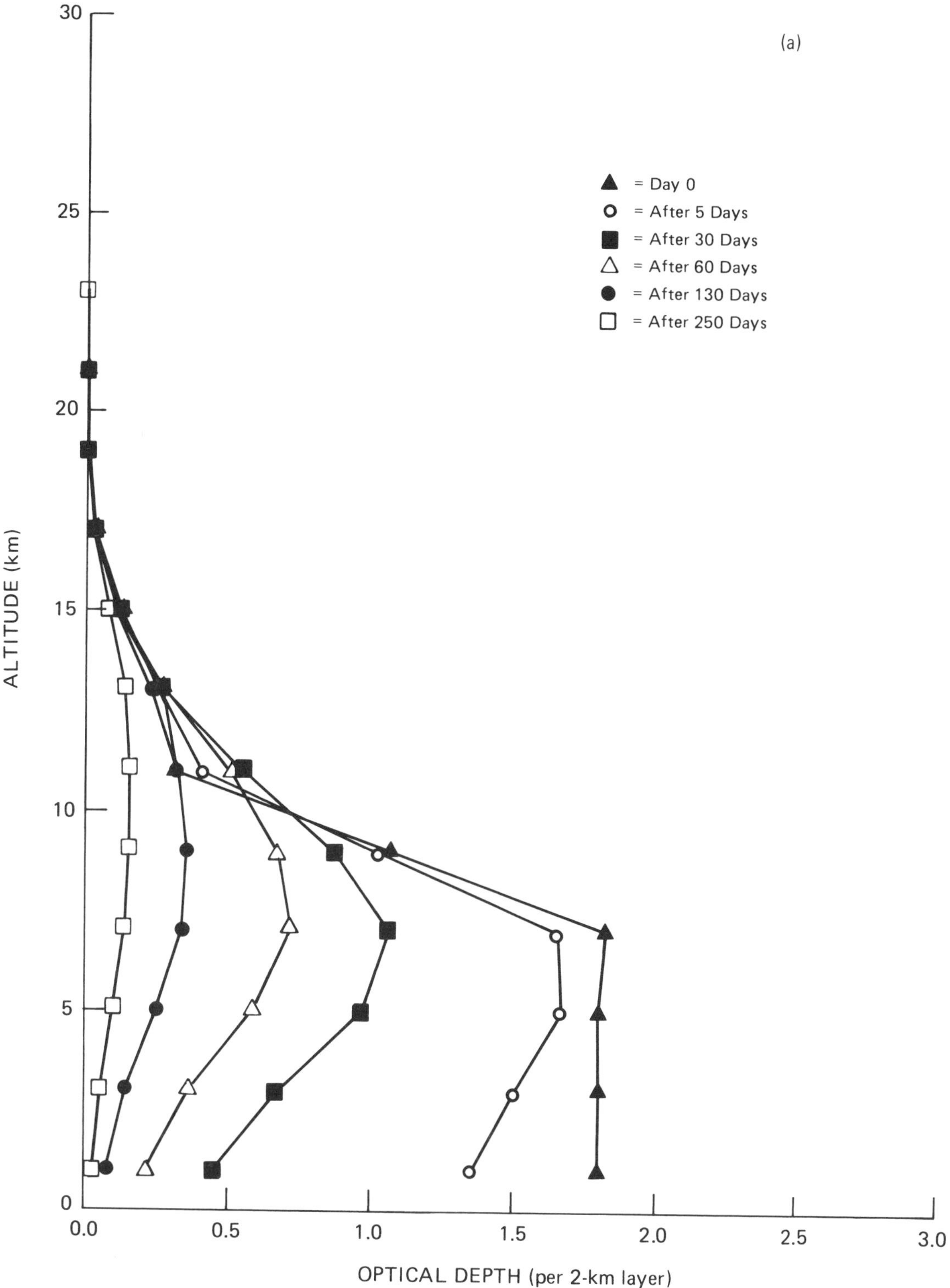

FIGURE 7.4 Contributions to total extinction optical depth for 2-km-thick layers at various times after hypothesized nuclear war smoke and dust injections. (a) NRC baseline, 30°N to 70°N; (b) NRC baseline, 0° to 90°N, fast rainout; (c) TTAPS baseline.

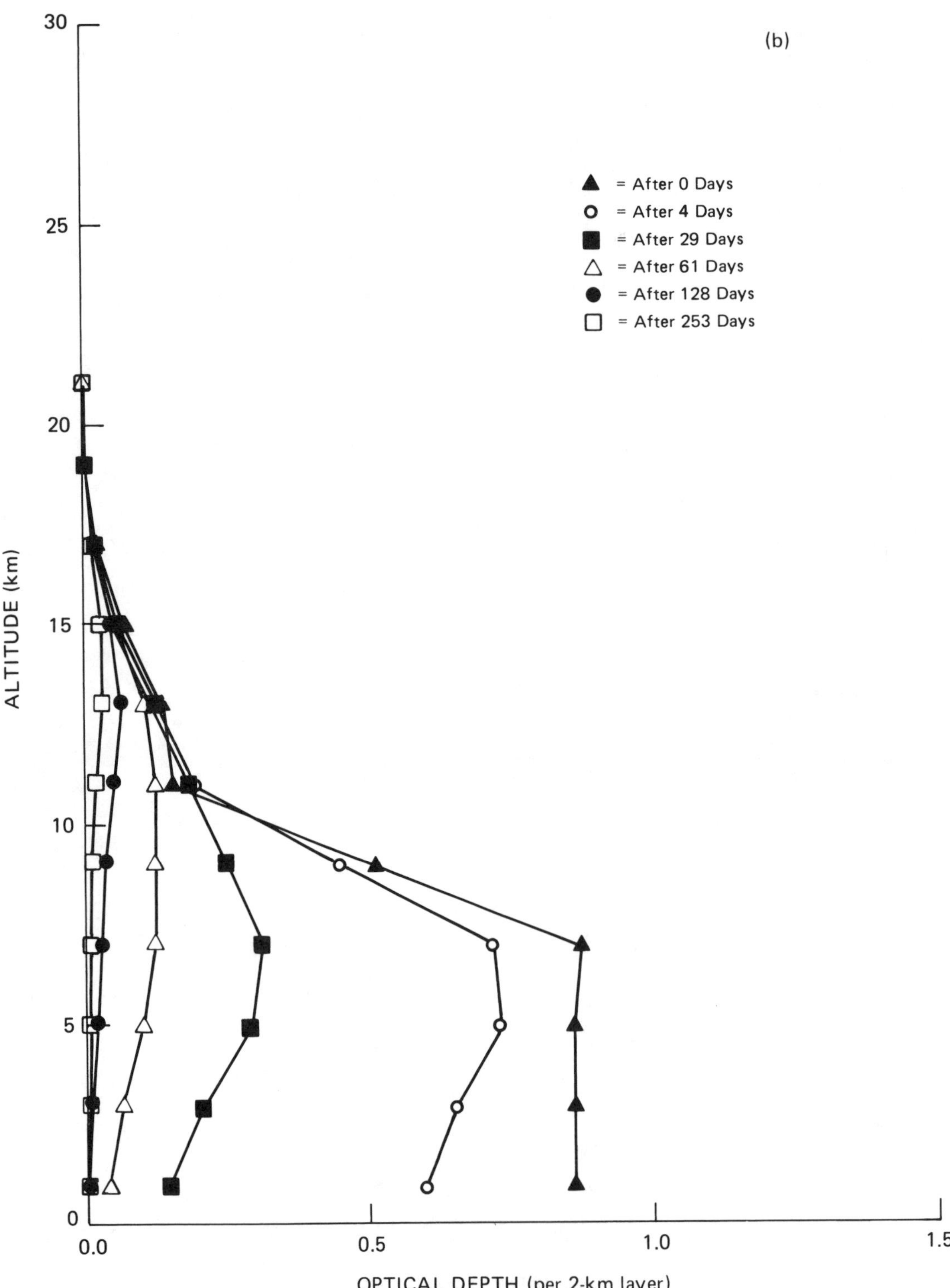

FIGURE 7.4 (continued)

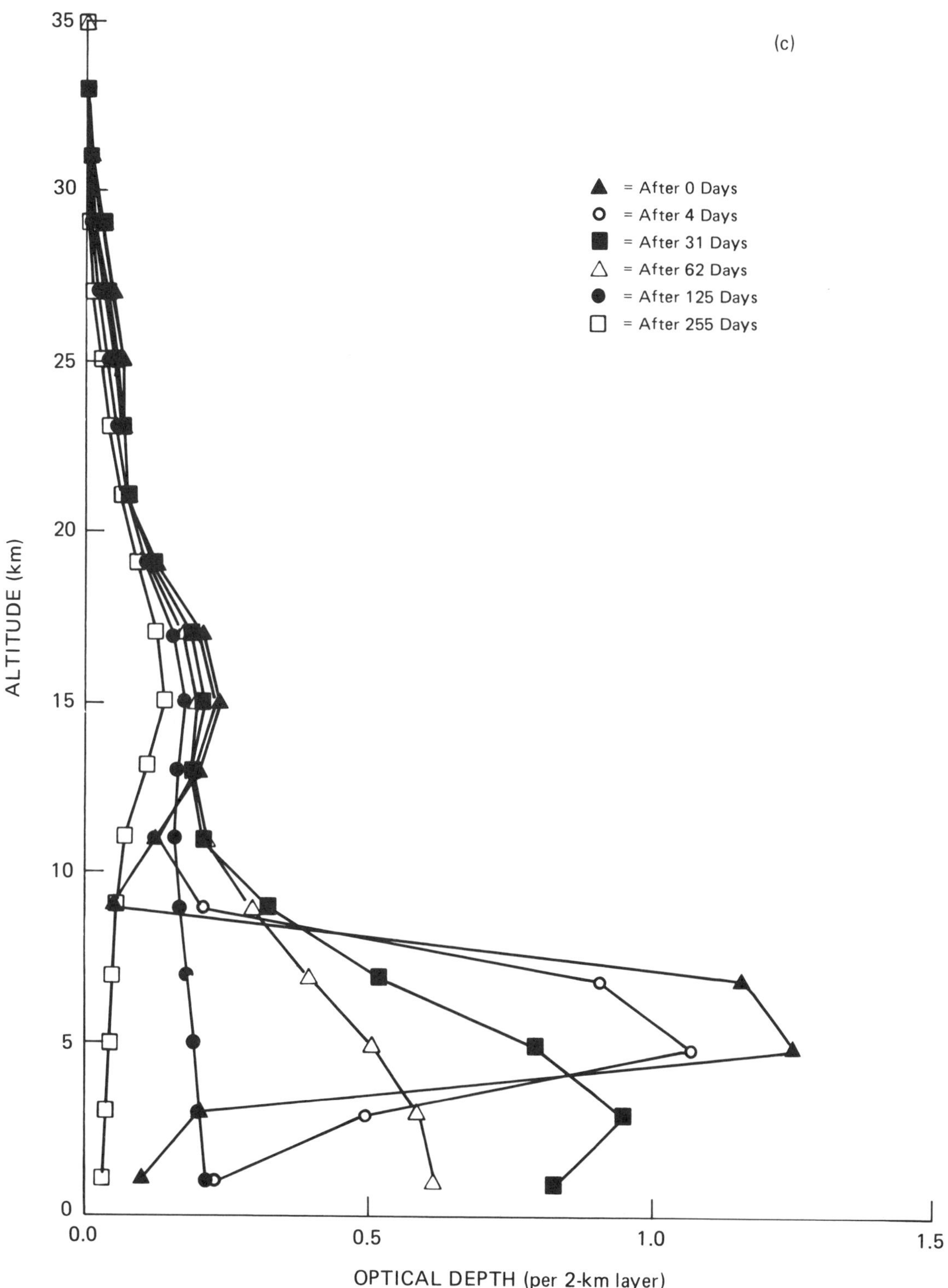

FIGURE 7.4 (continued)

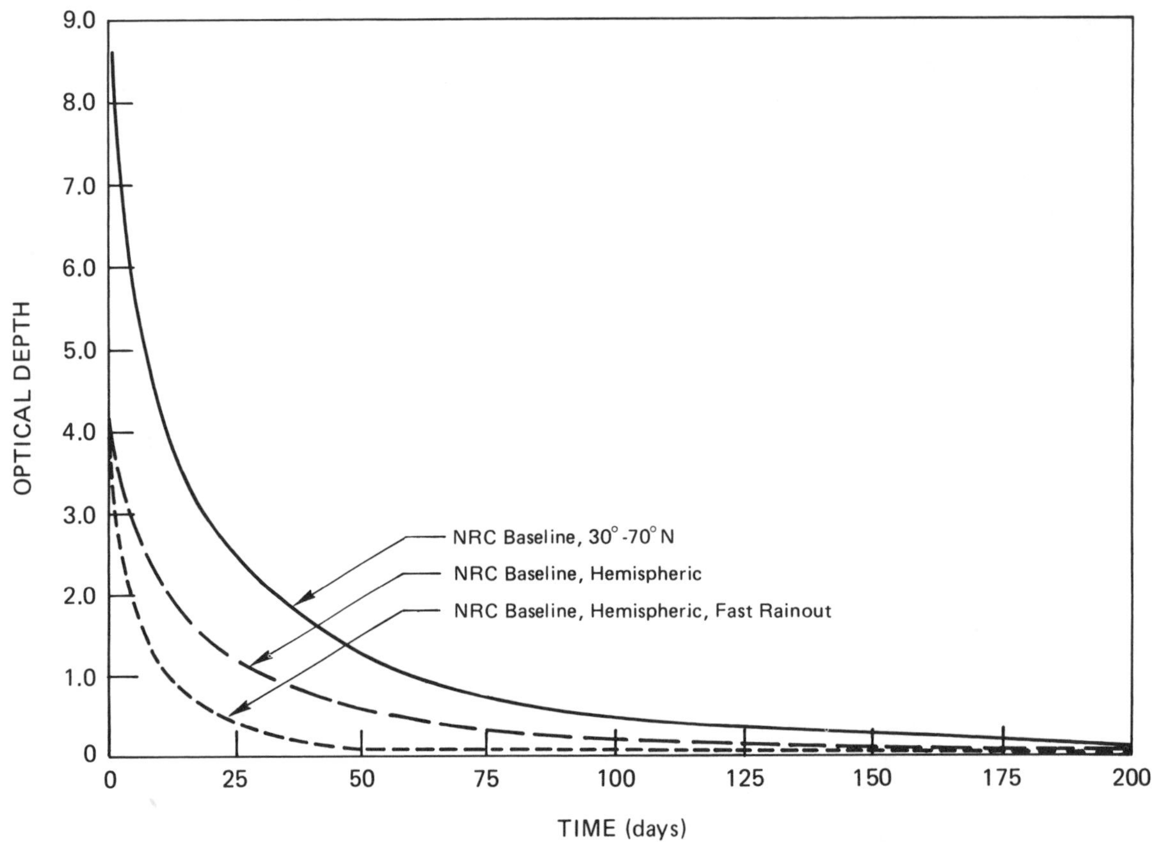

FIGURE 7.5 Time evolution of the total column extinction optical depth for the NRC baseline (30°N to 70°N), the NRC baseline injections (0° to 90°N), and the NRC fast-rainout variant (0° to 90°N). Contributions from both dust and smoke are included in the total.

"surface" temperatures given by the TTAPS model actually correspond to temperatures of a 2-km-deep lowest atmospheric layer. Because this has a significant thermal inertia, the neglect of land surface thermal inertia is probably not serious. Vertical mixing of heat occurs in the model only by convective adjustment.

Figure 7.6 shows the time evolution of near-surface temperature (actually the mean temperature of the lowest 2-km-thick atmospheric layer) for the NRC baseline, the excursion from the NRC baseline in which the smoke and dust clouds are distributed over the entire hemisphere rather than the 30°N to 70°N latitude strip, and the fast-rainout excursion from the NRC baseline. These temperatures drop rapidly to minima that are reached at times ranging from 10 to 25 days after the initial nuclear conflagrations. In each case there is a slower recovery, with the point of 50 percent temperature recovery reached at times ranging from 20 to 75 days.

These results are summarized in Table 7.3. For comparison, Table 7.3 also shows the results of several other calculations. The "no-fires" case from TTAPS (Turco et al., 1983b) isolates the effect of

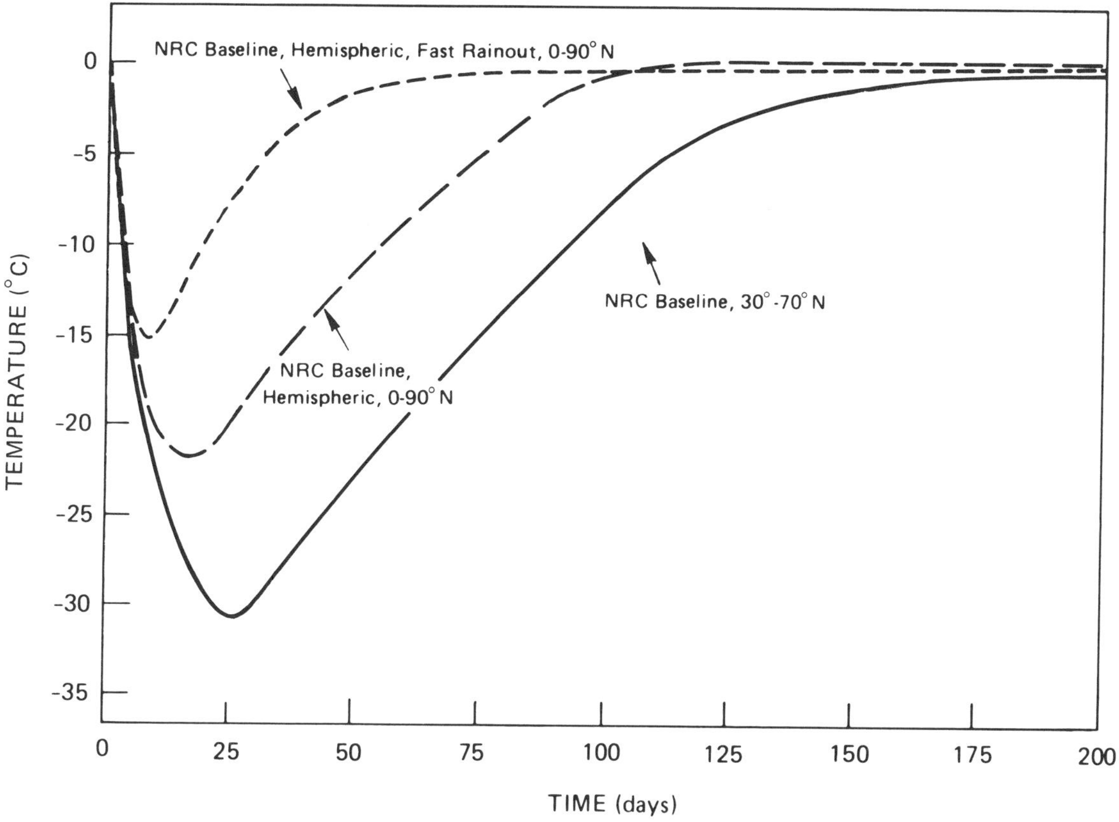

FIGURE 7.6 Time evolution of the surface temperature for the NRC baseline (30°N to 70°N), NRC baseline (0° to 90°N), and NRC fast-rainout variant (0° to 90°N). The temperature shown is actually the mean for the lowest 2-km atmospheric layer in the TTAPS model.

stratospheric dust alone, and has an initial dust optical depth of 1.4, so that it is within the bounds of the NRC 8500-Mt excursion for stratospheric dust discussed in Chapter 4. The LLNL one-dimensional model closely parallels the TTAPS one-dimensional model calculation (MacCracken, 1983). Initial injections include 207 Tg of soot from urban and forest fires, and 118 Tg stratospheric dust. Also included are the optical effects from the injection of nitrogen oxides into the stratosphere, taken as 8.3 Tg N, together with the corresponding ozone reduction. The net effect is a slight addition to the absorption of solar radiation in the lower stratosphere. Results from this calculation are similar to those of Turco et al. (1983b). Differences arise in part from differences in the treatments of precipitation scavenging and in the assumed optical properties of the particulates. The one-dimensional models do not provide reliable estimates of mean, or even typical, temperature changes over continental areas. Better estimates are provided by multidimensional models, also summarized in Table 7.3, though these too are subject to large uncertainties. Results from these models will be discussed in detail below.

TABLE 7.3 Surface Temperature Changes

Case[a]	Model Type	Region[b]	Season	Maximum Surface Temperature Change (°C)	Time of Temperature Minimum (days)	Time for Half Recovery (days)
NRC baseline, $\tau_o \sim 8$	1-D TTAPS	30°-70°N land areas only	Global average[c]	-31	25	76
NRC baseline hemis-pheric, $\tau_o \sim 4$	1-D TTAPS	0°-90°N land areas only	Global average	-21	17	51
NRC fast rainout, $\tau_o \sim 4$	1-D TTAPS	0°-90°N land areas only	Global average	-15	8	26
TTAPS baseline, $\tau_o \sim 4$	1-D TTAPS	0°-90°N land areas only	Global average	-37	28	76

TTAPS slow rainout, $\tau_o \sim 6$	1-D TTAPS	0°-90°N land areas only	Global average	-42	35	120
TTAPS no fires, $\tau_o \sim 1.4$ due to strato-spheric dust	1-D TTAPS	0°-90°N land areas only	Global average	-14	47	~300
LLNL, (similar to TTAPS baseline[d]), $\tau_o \sim 5$	1-D LLNL	0°-90°N land areas only	Global average	-32	14	47
LLNL, (similar to TTAPS baseline[e])	2-D LLNL	Land areas only at 30°N	Global average	-16	~10[f]	70
LLNL, (similar to TTAPS baseline[e])		30°-60°N land areas only	Global average	-11	~10[f]	70

TABLE 7.3 (continued)

Case[a]	Model Type	Region[b]	Season	Maximum Surface Temperature Change (°C)	Time of Temperature Minimum (days)	Time for Half Recovery (days)
NCAR, NRC baseline	3-D GCM NCAR-CCM)	30°-60°N land areas only	Summer	-26[g]	~10[f]	--
NCAR, NRC baseline	3-D GCM NCAR-CCM)	30°-60°N land areas only	Spring	-17[g]	~10[f]	--
TTAPS baseline	EBCM Robock	30°-60°N land areas only	Summer	-17	30-60	~100
TTAPS baseline	EBCM Robock	30°-60°N land areas only	Spring	-14	60-90	150-200

[a]Cases are listed according to the injection scenario used; τ_o is the initial estimation of optical depth.

[b]Except for the LLNL two-dimensional model calculation, the regions given correspond closely to both the continental portion of the region covered by smoke and dust and the region over which the temperature perturbation has been averaged. For the LLNL two-dimensional calculation, dust and smoke were assumed to be injected nonuniformly in the 30°N to 60°N belt; the region given in this table corresponds to the region over which the temperature perturbation has been averaged.

[c]"Global average" refers to globally averaged insolation. Conditions correspond most closely to 30° latitude at equinox, but because diurnal variations are not modeled, the correspondence is not exact.

[d]See page 143 for details.

[e]See page 151 for details.

[f]For these cases, temperatures were picked off of the results for day 10. They correspond closely but not precisely to the times and values of the maximum surface temperature depressions.

[g]These values are the average maximum temperature changes for the 30°N to 60°N land areas. They do not correspond to the extremes within those areas.

NOTE: The one-dimensional results are given to illustrate sensitivities only. Temperature changes given by these one-dimensional models do not indicate mean or even typical continental values (see page 150).

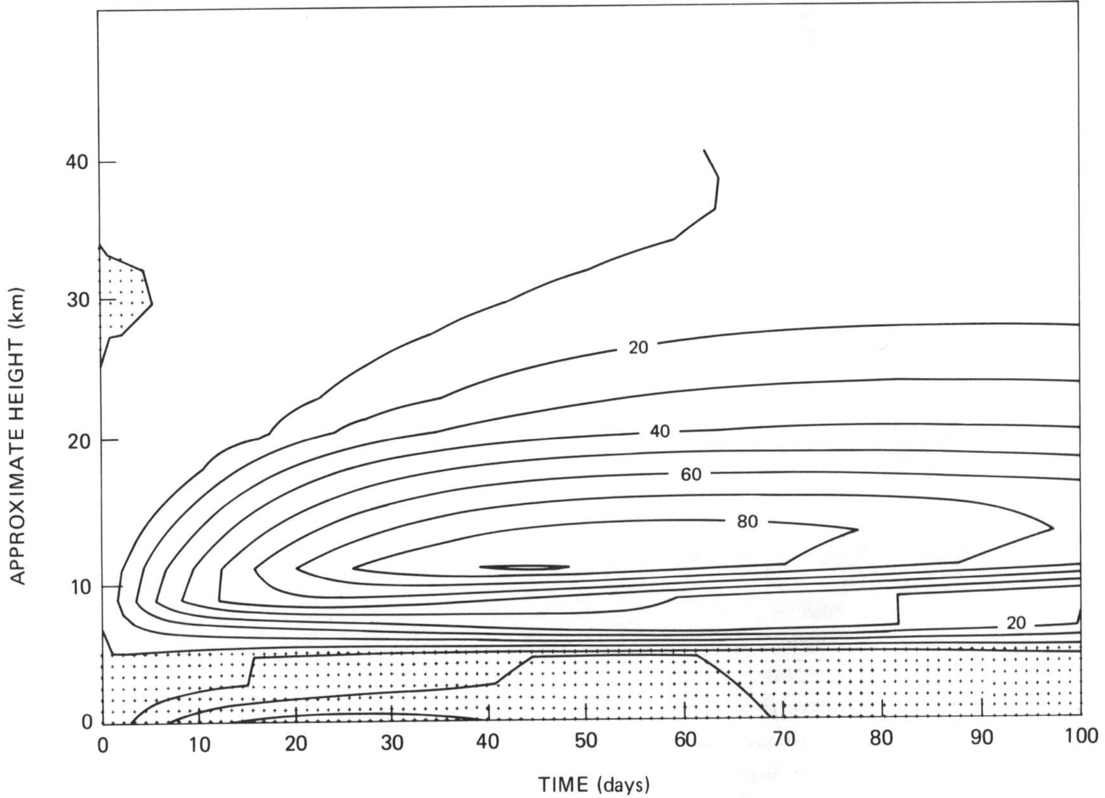

FIGURE 7.7 Time-height section of the temperature perturbation (°C) for the NRC baseline case (30°N to 70°N). The stippled area represents negative values.

The time evolution of the vertical temperature change profile for the NRC baseline is shown in Figure 7.7. As with the cases considered by the TTAPS group, the strongly absorbing particulate clouds produce rapid increases in temperature aloft, leading to a radical modification of the initial temperature profile, and replacing the normal tropospheric lapse rate with a deep and intense inversion. As a consequence of the optical saturation effect mentioned above, one-dimensional model responses are not sensitive to the precise amount of smoke injected, provided the optical depth is in the neighborhood of 2 or greater--nor is the magnitude of the temperature increase sensitive to the details of the vertical distribution of smoke provided that substantial amounts of smoke remain above about 4 km. Except for the NRC hemispheric fast-rainout excursion, in which the optical depth drops below 2 within 7 days, each of these cases shows a maximum temperature increase near 80 K. To illustrate this insensitivity, magnitudes, times, and altitudes of maximum temperature increases for several cases are given in Table 7.4.

The essence of the temperature change tendencies indicated by the numbers in Tables 7.3 and 7.4 derives from the efficiency of the smoke particles as absorbers of solar radiation coupled with their inefficiency for absorption and emission of infrared radiation. The

TABLE 7.4 Elevated Layer Temperature Changes

Case	Region	Season	Maximum Temperature Increase (°C)	Time of Maximum (days)	Height of Maximum (km)
NRC baseline	30°-70°N	Annual average	+85	43	11
NRC baseline hemispheric	0°-90°N	Annual average	+65	47	11
NRC hemispheric fast rainout	0°-90°N	Annual average	+35	30	11
TTAPS baseline	0°-90°N	Annual average	+95	55	17

ratio of the extinction efficiency of the smoke at a wavelength of 0.55 μm to that at 10 μm is about 10 for the NRC baseline case. This ratio follows from the model of the properties of the injected smoke in the baseline model discussed in Chapter 5. The ratio of absorption efficiencies is about 4. As long as the absorptivity of the cloud for solar radiation is sufficiently large that the altitude at which most of the solar radiation is absorbed is higher than the altitude from which most of the thermal infrared radiation is emitted, the normal warming of the surface due to the "greenhouse effect" will be cut off, and the "antigreenhouse" tendencies depicted in Figures 7.6 and 7.7 will appear in radiative convective calculations. This conclusion remains valid even for optical depths at visible wavelengths that are sufficiently large that the infrared opacity is also large, so long as the altitude of solar energy absorption is above the level of thermal infrared emission (T.P. Ackerman, private communication, 1984; Golitsyn and Ginsburg, 1984). In the limiting case of a globally uniform absorbing cloud in thermal equilibrium, with solar radiation absorbed above the altitude of thermal emission to space, the mean surface temperature would correspond approximately to the effective emission temperature of a planet in thermal equilibrium with the incident sunlight (about 255 K for the present earth) while the cloud temperature would be much warmer, the exact temperature depending on cloud thermal emissivity.

Because of the optical depth saturation effect for smoke absorption discussed on pages 132 and 133, the one-dimensional model temperature changes are insensitive to changes from the NRC baseline in smoke particle size distribution as optical properties. For the same reason, the results are not very sensitive to the initial mass of smoke input, provided the initial mass is large enough. Lower injected smoke mass (or smoke absorptivity) by a factor of about 4 would lie near the edge of the saturation regime and climatic effects would decrease rapidly for smoke injections in this range and below. In addition, the maximum temperature depression is sensitive to smoke lifetime if rainout rates are unexpectedly large, as in the NRC fast-rainout case. In this case, the smoke opacity may not remain above the strong absorbing threshold (optical depth of about 2) long enough for even approximate thermal equilibration of perturbed continental surface temperatures to take place.

THERMAL AND CIRCULATION EFFECTS CALCULATED BY MULTIDIMENSIONAL MODELS

The one-dimensional models discussed in the previous sections should not be expected to predict accurate temperature changes, especially near the surface. They are useful for evaluating the possible magnitude of the problem, and for efficiently examining the sensitivity of the thermal tendencies to various assumptions. As mentioned above, the temperature changes predicted by these models should not be interpreted as mean or even typical changes for the regions given in the third column of Table 7.3. These models might provide upper limit values for the given smoke and dust inputs and removal rate assumptions, applicable perhaps to the deep interiors of continents where the ameliorating influences of the oceans do not apply. Even for this limited quantitative purpose, such results should be considered with some reservations, however, because patches with much larger optical depth might prevail for some time over localized regions of the continental interiors, and might lead to larger temperature drops.

There are three serious limitations to the use of one-dimensional models to deduce representative temperature changes. The first problem is that they neglect the ameliorating effects of the nearly constant sea-surface temperatures on the climate over coastal areas and continental areas exposed to fresh flows of maritime air. This limitation would be most serious for the near-surface temperature changes, because the temperature increases due to absorption of solar radiation aloft would be likely to develop and persist over the oceans as well as over continents. The second problem is somewhat more subtle. As pointed out by MacCracken (1983), the nonuniform distribution of dust implies that some regions would have very high optical depths while others would be nearly clear (note the patchiness in Figure 7.1, for example). Thus fewer smoke particles would be exposed to sunlight than would be the case if the same amounts of smoke and dust were uniformly distributed, and as a consequence the absorptive power of the particulates would be less efficiently utilized; the uniform optical depth assumption tends to maximize the overall mean thermal tendencies. A third problem is that these models do not predict or account for changes in water cloud distributions, which could in turn have an additional effect on the radiation balance. That this effect could be important is illustrated by a one-dimensional model calculation of MacCracken (1983), in which a 33 percent higher induced surface temperature drop occurs for a case without water clouds than for the corresponding case with normal clouds.

In one attempt to account for these effects, MacCracken has used a two-dimensional meridional plane climate model to calculate temperature changes for particulate optical depths specified as meridional averages obtained from the Oregon State GCM (Gates and Schlesinger, 1977). Transport and removal processes in this calculation were accounted for using the LLNL two-dimensional model (this model used the LLNL "GRANTOUR" code; cf. Walton and MacCracken (1984)). The initial particulate injections were, as mentioned above, 118 Tg of dust and 207 Tg of smoke. The circulation responsible for particulate transport was represented by only a single mid-troposphere wind, but the climate model included algorithms for predicting meridional circulation, cloudiness, precipitation, and the hydrological cycle based on weighted means of land and ocean over each latitude belt (MacCracken, 1983). The result is summarized in Table 7.3 for the normal rainout parameterization of the LLNL two-dimensional model, calibrated to represent the unperturbed atmosphere.

The maximum surface temperature decrease for this model was much less than for the one-dimensional LLNL model. This difference is primarily due to the treatment of the exchange of heat between continental and oceanic regions and to the more realistic nonuniform and evolving spatial distribution of smoke in the two-dimensional case. The calculated 30°N value should not be taken to represent the extreme temperature perturbation that might occur at mid-latitude continental locations with this scenario; the extreme may be substantially larger. MacCracken has also carried out a two-dimensional calculation with the rainout removal rate suppressed to simulate the possible effects of the nuclear war smoke and dust injections on precipitation, and in this case somewhat larger and much more persistent temperature perturbations developed.

A GCM calculation has been carried out by Covey et al. (1984) using the NCAR Community Climate Model (CCM; Washington, 1982). The horizontal resolution of this model corresponds to approximately 4.5° latitude and 7.5° longitude, and the model has nine layers in the vertical. Other details are given in Covey et al. and references therein. Smoke was introduced as a perturbation to initial atmospheric states corresponding to the undisturbed climate for winter, spring, and summer seasons. The total amount of smoke, its distribution, and its absorptivity for solar radiation corresponded closely to the NRC baseline case: nearly uniform concentration contained in the 27°N to 71°N latitude and 0 to 10 km altitude band, with an absorption optical depth of 3.0. Dust injection was neglected. Multiple scattering and infrared absorption and emission by the smoke particles was also neglected, and the smoke distribution was not allowed to change with time. The NCAR CCM accounts for absorption and emission by atmospheric gases at all wavelengths and scattering and absorption by water clouds. The distribution of water clouds (liquid or ice) is calculated based on relative humidity distributions and convection predicted within the model.

Zonally averaged temperatures for the summer case averaged over the period from 10 to 20 days after smoke injections are shown in Figure 7.8. There is a large elevated temperature maximum that is much like that predicted by the one-dimensional models, except that it spreads in an attenuated form over a far wider latitude range than that of the smoke. The continental average surface temperature change 10 days after smoke injection averaged over the 30°N to 60°N latitude band was -26°C for the summer case and -17°C for spring (Table 7.3; Thompson et al., 1984). Temperature changes at the surface averaged over days 6 to 10 following the nuclear attacks are shown in Figure 7.9. By day 10, in this case, subfreezing temperatures extend over much of the Eurasian and North American continents, although the western part of Eurasia and coastal strips of North America are not so seriously affected. As Covey et al. point out, this calculation also shows the rapidity with which subfreezing temperatures might develop under large patches at relatively low latitude. This model predicts a number of meteorological effects in addition to temperature change, as discussed in the next section.

Another GCM calculation has been carried out at the Computing Center of the USSR Academy of Sciences by Aleksandrov and Stenchikov (1983), using a modified version of the Oregon State two-level model. The calculation starts from simulated normal annual mean conditions with clouds of absorbing particulates injected into both the troposhpere and the stratosphere. In this calculation, the particulates are assumed to be distributed uniformly over the entire northern hemisphere poleward of 12°N. Only absorption of solar radiation is taken into account, and the initial optical depth for absorption is about 7, so that, if one allows for the areal extent of the cloud, the initial mass of material corresponds to about 4 times that of the NRC baseline case. The smoke lifetime is also assumed to be rather large with the optical depth remaining above 3 until day 100.

This calculation produces a decrease in zonal mean surface temperature of as much as 22°C at 65°N on day 40, and a strong temperature inversion throughout the northern hemisphere. Extreme surface temperature changes produced by the model are quite localized, with extreme departures exceeding -40°C by day 40 in patches throughout the land-covered area of the northern hemisphere. Based on the one- and two-dimensional model results discussed previously, these results on day 40 are reasonable, given the very large initial injections and slow removal rates of this calculation. The initial zonal mean state and the temperature distribution after disappearance of the smoke and dust clouds at 360 days show unrealistic features, however, possibly due to the low horizontal and vertical resolutions of the model used for this early investigation of the problem. Discussion, interpretations, and comparisons of the results of the CCM and the USSR Academy of Sciences models are given in Thompson et al. (1984).

In comparing the one-, two-, and three-dimensional model simulations for similar scenarios and taking seasonal effects into account, the multidimensional models give average continental surface coolings that are smaller than those given by the one-dimensional models by factors of

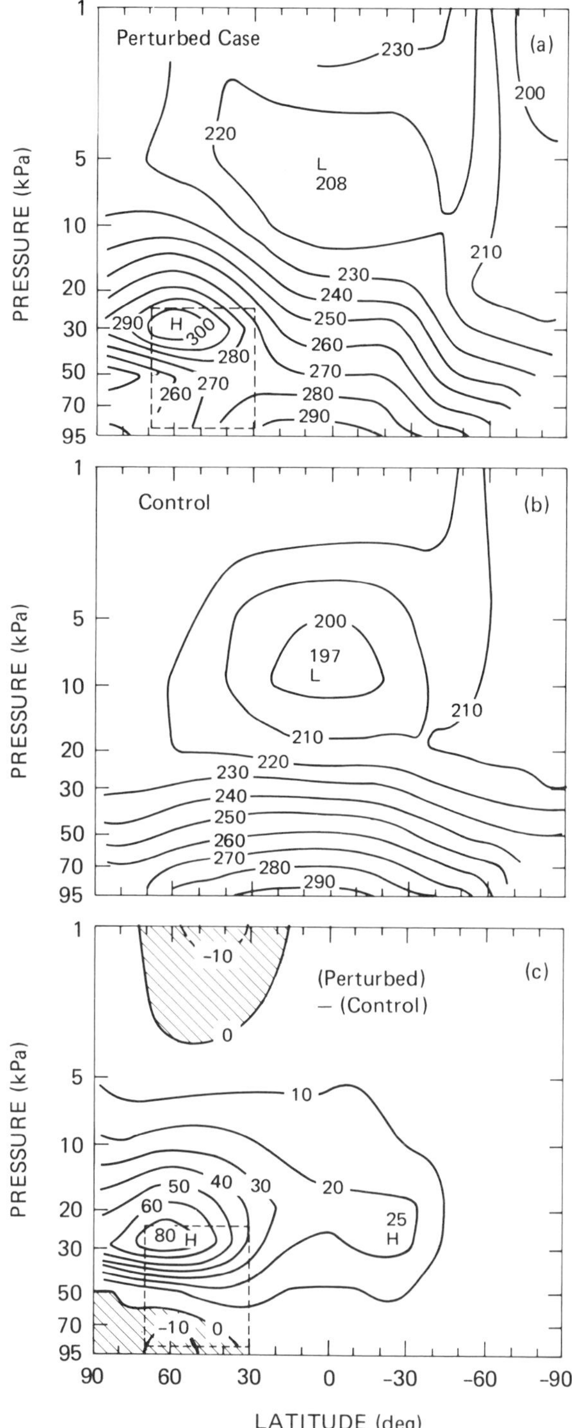

FIGURE 7.8 Meridional temperature cross-section for the perturbed case, the control case, and the perturbed minus the control at t = 10 days from the CCM calculations for summer. The vertical scale is pressure, with 10 kPa corresponding approximately to 15 km and 7 kPa to 30 km. (From Covey et al., 1984.)

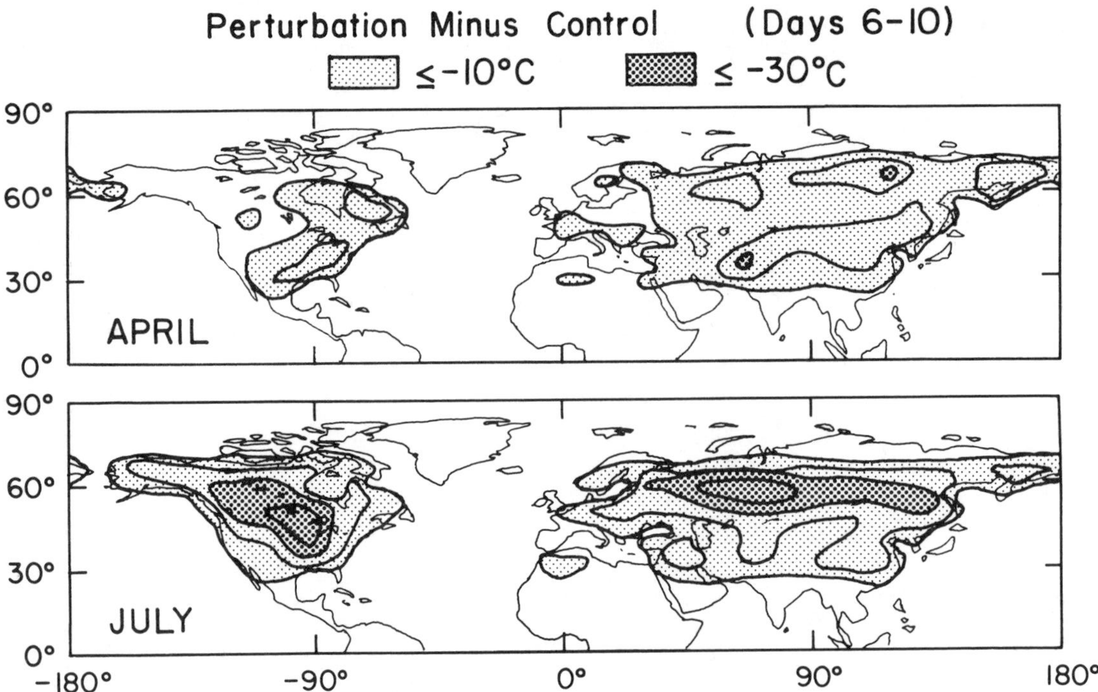

FIGURE 7.9 Surface temperature change from the control after 10 days for a summer run of the CCM. (From Thompson et al., 1984).

2 to 3. This difference, which was anticipated by the TTAPS group, can be understood primarily as a consequence of the exchange of heat between ocean and continent as it is represented in the two- and three-dimensional models; in addition, the evolving smoke distribution plays a role in MacCracken's two-dimensional calculation. The extremes of mid-continent surface temperature decrease found in the NCAR CCM calculations are comparable to those occurring in the one-dimensional calculations, and this is consistent with what one would expect given the limitations of the one-dimensional models, for scenarios in which dense clouds of smoke remain over the continental interiors for the order of 10 days or more.

It is worthwhile to note the possible effects of some of the improvements in this type of calculation that future research should address. The inclusion of scattering of sunlight and infrared absorption and emission, including the optical effects of injections of dust, would produce different predicted temperature changes. Multiple scattering and inclusion of dust would probably lead to somewhat lower near-surface temperatures, since the particulate cloud would have a higher albedo than the cloud assumed in the CCM simulation.

The initial cooling of continental areas is faster in the CCM calculation of Covey et al. than in the calculations using the TTAPS one-dimensional model. This difference is almost certainly due to differences in the treatment of near-surface thermal inertia and vertical resolution. The "surface" temperature in the TTAPS model corresponds to the lowest 2 km of the atmosphere (see page 142). As a result, this model does not allow for the possibility of very shallow cold layers over dry continental surfaces that form at early times in the CCM calculation. Accurate treatment of the near-surface boundary layer is a prerequisite for reliable estimation of surface temperature changes.

The neglect of infrared absorption and emission due to the smoke in the CCM calculations could lead to exaggerated cooling rates, but detailed calculations show that this effect is not important at early times for the NRC baseline injections (Ramaswamy and Kiehl, 1984). On the other hand, it could be significant during winter at middle and high latitudes, since the contribution of thermal infrared radiation to the surface energy balance is more important relative to solar radiation during winter. Infrared emission could also significantly affect cooling rates if smoke particle sizes are much larger than those of the NRC baseline.

Simulations that permit the transport of particulates should have a high priority in future research. Such models allow calculation of the early spread of smoke and dust clouds and explicit calculation of the feedback loop between the particulates and circulation; not only would transport spread the cloud, but the enhanced circulation driven by the heating produced by the cloud could accelerate the spread.

Rainout and aging of smoke particles could have significant ameliorative effects on the thermal perturbation after 10 to 20 days. Too little is currently known about fundamental aspects of these processes to permit convincing modeling at present, although parameterized treatments of rainout could be incorporated with relative ease. In addition, diurnal variations should be explicitly included in future studies, not only because the diurnal near-surface temperature extrèmes are of practical importance, but also because diurnal heating and temperature variations would be associated with diurnal variations of wind and vertical mixing that could be quite important for particulate transport and removal. Diurnal variations should even be included in future studies involving one-dimensional radiative-convective model calculations. The climatic impact of smoke and dust injections is slightly smaller when diurnal variations are explicitly simulated in these models (Cess, 1984).

Finally, seasonal variations need to be considered in more detail. The one-dimensional models have, at the time of writing, tended to focus on annual mean conditions, as did Alexandrov and Stenchikov. Covey et al. carried out a calculation for winter using the CCM, but did not report in detail on the results. Robock (1984) has used an EBCM to simulate nuclear exchanges in autumn and winter as well as spring and summer. However, in this calculation, as in the CCM calculation, infrared radiation absorbed and emitted by the particulates was neglected. Because of the relatively important role played by thermal infrared radiation during winter, the significance of these results is not clear. For this reason, as well as because winter is the dormant season in middle and high latitudes, the ecological effects of the smoke and dust clouds could be less serious during winter. Winter cases deserve serious investigation, however, because a postwar population already exposed to the rigors of winter could be particularly sensitive to the additional darkness and persistent cold that could be produced by nuclear smoke and dust clouds. Moreover, transport and dispersion rates are normally greater during winter, so that tropical and subtropical regions whose ecologies are particularly sensitive to meteorological excursions could be at greater risk (but see the discussion in the next section of the possibility of greatly enhanced meridional transport during summer and spring).

MODIFICATION OF CLOUDINESS, PRECIPITATION, AND WINDS

The strong thermal effects indicated by studies reviewed in preceding sections would certainly produce large changes in other climatically significant quantities. The thermal effects are themselves uncertain, and deductions about any consequent effects are necessarily even less certain, and must be considered somewhat speculative. Nevertheless, in this section an attempt is made to assess such effects, drawing on analogies to known meteorological phenomena and, where possible, on available model results. The following phenomena are considered: fog, cloudiness and precipitation distributions, zonal mean winds, other large-scale wind systems, and ultra-high clouds.

Ground Fog

Under the influence of large-scale dust and smoke clouds, radiation fogs would form over land areas as the surface temperatures dropped below the dewpoint. Initially at least, these fogs could provide some protection against further temperature decreases, particularly in affected tropical or subtropical regions where dewpoints are normally high.

The lifetimes of such induced radiation fogs and the amount of thermal protection they would provide are uncertain. Normally, when radiation fog is persistent over a period of days, there is a tendency for thermal balance, in the diurnal mean, between the competing effects of cooling of the entire foggy layer by emission of infrared radiation, heating due to that portion of the incident sunlight not reflected by the fog and therefore absorbed at the ground or in the layer, and entrainment of warm air from above.*

Such fogs are usually most persistent under otherwise clear skies during winter with strong subsidence above the foggy layer and near-stagnation of the low-level winds. If some mixing occurs in the surface layer, the fog can lift to form a low stratus layer. Such fog and stratus layers have been known to persist for as long as a month in California's Central Valley. Under postnuclear war smoke and dust clouds, sunlight would be virtually absent, and ventilation conditions would probably be at least as variable as in the unperturbed

---

*While fogs are forming, they release latent heat, but this effect would be small in comparison with the radiative perturbations of the nuclear war scenarios.

atmosphere. Consequently, one could expect that fog or low stratus layers would persist in some places, but would be rapidly removed in others; and that, where persistent fog did occur, it would be somewhat less effective in ameliorating surface temperatures than similar fogs for the unperturbed atmosphere.

There is little hard evidence bearing on this question. The CCM calculations of Covey et al. and the two-dimensional model results of MacCracken do show increases in low-level cloudiness in the perturbed regime, but improved boundary layer formulations are needed for a more precise assessment of the role of fog. Because such fogs may be persistent and have ameliorative effects, the sensitivity of surface temperature changes over extensive continental areas to particulate lifetime could be more pronounced than indicated by the considerations of the preceding sections.

Cloudiness and Precipitation

The heated elevated layer produced by absorption of sunlight in a widespread smoke cloud would suppress convection and prevent the formation of clouds within the layer because of the increase in static stability and lowering of relative humidity associated with high temperatures in the layer. Since precipitation normally forms only in deep clouds, it would be suppressed as well, at least over the continents.

The situation over the oceanic regions covered by particulate clouds would be quite different from that over the continents. While the continental boundary layer would be statically stable, the oceanic boundary layer would generally be unstable, at least in regions affected by flow of air from the cold continents over the still-warm surface waters. As a result, enhanced convection would be likely over large areas of the ocean, especially adjacent to continents. Because of the high-temperature layer aloft, this convection would probably be shallow, however, and would perhaps be similar to that observed when very stable cold polar air from the continents flows over adjacent oceans. These convective boundary layers are rarely more than 2 km deep (Walter, 1980). Clouds of this depth can produce showers, however, especially if temperatures are subfreezing below the cloud top level. Thus such oceanic convective boundary layers could be regions of frequent precipitation and effective particulate removal. Coastal regions might also be regions of enhanced cyclonic storm activity because of the increased temperature gradient between land and sea. This effect could add to the effectiveness of these areas as sites of particulate removal. However, if there were a well-developed elevated heated layer, it is unlikely that even the cloudiness associated with such coastal storms could penetrate to very great heights.

Convective cloudiness and precipitation could be enhanced near the edges of major smoke cloud bands, and in the vicinity of outlying streamers. Chen and Orville (1977) have shown that, under suitable conditions, small rain showers could be produced by seeding the atmosphere with carbon black particles. The effectiveness of precipitation developing along smoke cloud edges is extremely difficult to estimate. This process is likely to be most effective at early times when the cloud edge-to-area ratio is largest and before a widespread hot elevated layer has had a chance to develop. The possibility of enhanced precipitation during this stage of smoke cloud evolution constitutes one of the major uncertainties in the analysis. As Figure 7.8 shows, full development of the stabilizing warm layer extends far beyond the edges of the smoke cloud. This effect, which is due to the tendency of subsiding motion in the cloud environment to compensate rising motion in the cloud region, is likely to suppress deep convection near the cloud edges after a week or two.

Zonal Mean Winds

Changes in the zonally averaged west-to-east winds are relatively easy to calculate, given any set of temperature field changes, since wind changes will be in thermal wind balance* with the temperature changes. For the summer GCM simulation of a nuclear war scenario by Covey et al., the predicted changes in the mean west-to-east winds are shown in the bottom panel of Figure 7.10. These changes correspond to the temperature changes shown in the bottom panel of Figure 7.8. The main features are strongly enhanced west-to-east winds in the region normally occupied by the stratosphere at middle and high latitudes of the southern hemisphere and high latitudes of the northern hemisphere, and very strongly enhanced easterly winds above the subtropical (northern hemisphere) edge of the smoke. Each of these features can be understood in the light of the thermal wind relation as it applies to the temperature perturbation field. Their effects would be to greatly increase the rate of zonal transport of particulates, especially if some of the particles were to spread upward and equatorward over time.

There is more uncertainty associated with predictions of the change in the zonally averaged meridional component of the flow, but, given the altered heating field, predictions can be made with reasonable confidence (see Held and Hou, 1980, for a discussion of the theory of the zonally averaged circulation). For the spring and summer simulations of Covey et al., the nuclear-war-induced changes are dramatic. Figure 7.11 shows the stream function for the meridional circulation mass flow for the spring control and perturbed cases. Intense heating along the tropical edge of the cloud has caused the normal two-cell circulation to be replaced by a single cell with rising motion near the southern edge of the cloud. The intensified heating has also intensified and elevated the meridional flow toward the southern

---

*That is, conditions that are completely determined by the temperature and surface pressure distributions. This relationship is a consequence of the close balance between Coriolis and pressure gradient forces in large-scale terrestrial wind systems (geostrophic wind balance). It should be noted that the true temperature change field inevitably includes the effects of dynamical as well as radiative processes.

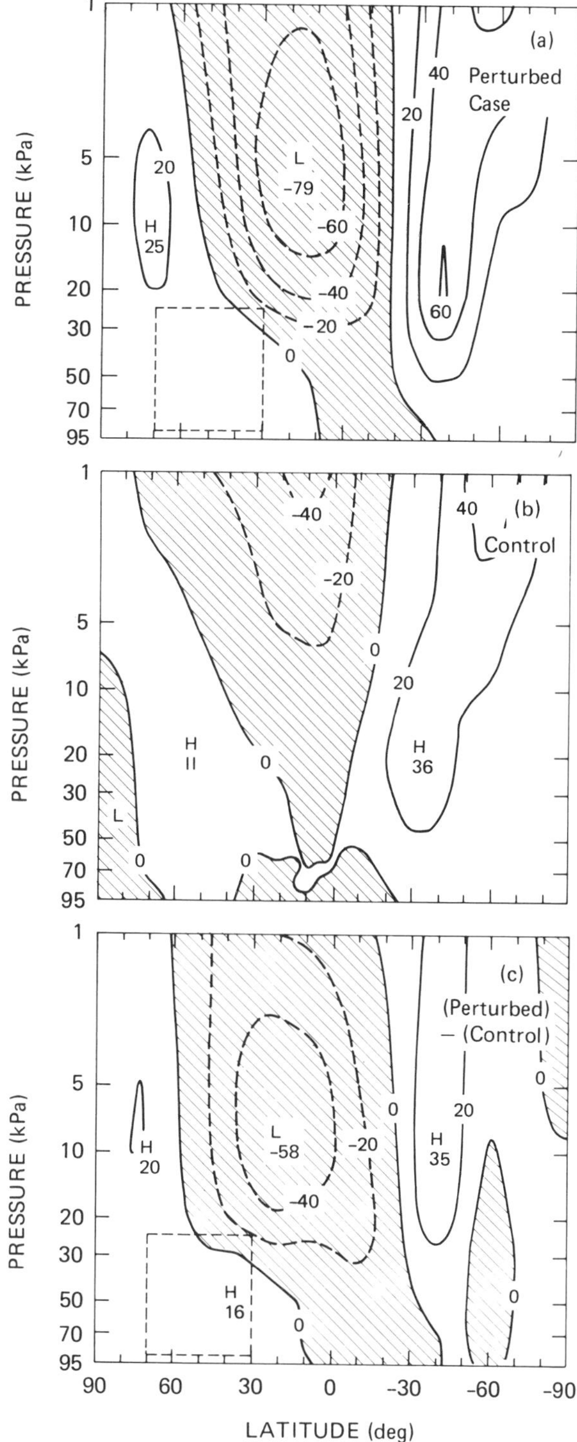

FIGURE 7.10 Zonal wind cross sections for the perturbed case, the control case, and the perturbed minus the control at t = 10 days from the CCM calculations for summer. Isolines are labeled in meters per second with positive values eastward; westward winds are shaded. (From Covey et al., 1984.)

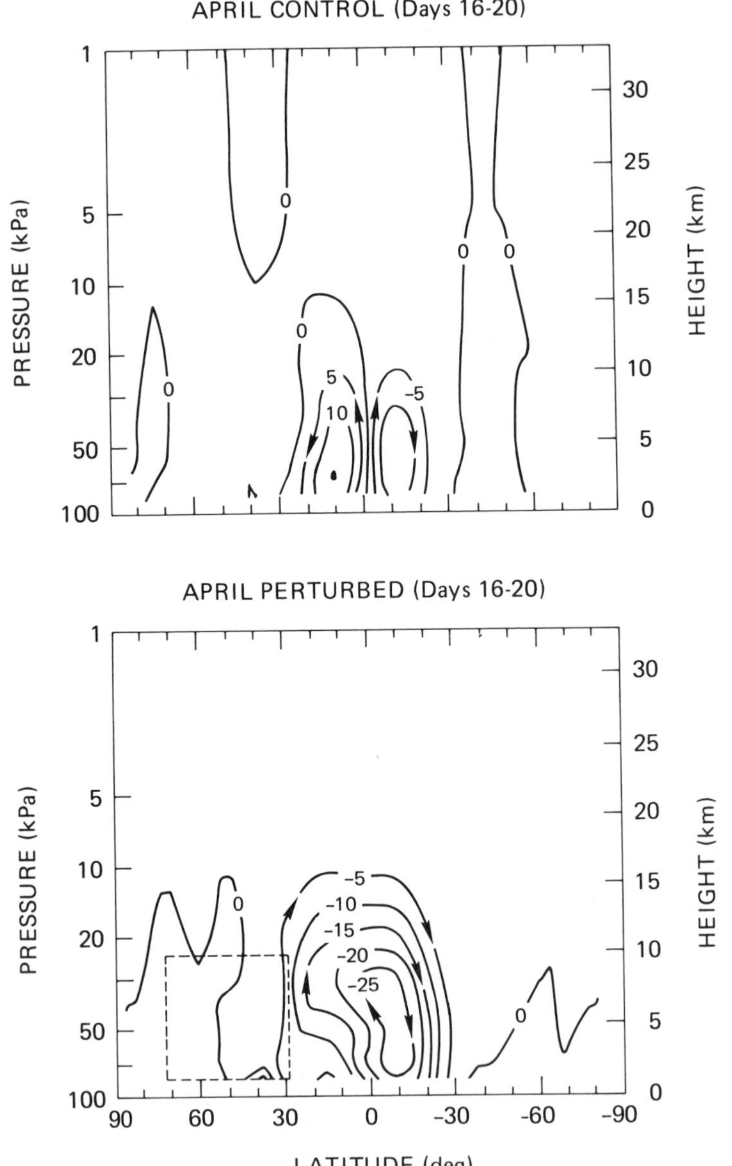

FIGURE 7.11 Meridional circulation mass flow stream functions from the NCAR CCM for the April (spring) control case and for the perturbed case. Between any two contours, the mass flow is 10^{10} kg/s. Averages over the period t = 16 to 20 days are shown. (From Covey et al., 1984.)

hemisphere. The mean meridional velocity in this branch is about 4 m/s. Note that the descending branch of the meridional cell spreads southward about 10° latitude as well. This intensification of the mean meridional circulation did not occur in the case of a perturbed circulation calculated by Covey et al. for winter because the differential heating along the southern edge of the cloud was not intense enough during winter to counter the normal circulation driven by heating centered farther south.

A strong enhancement and widening of the mean meridional circulation also developed in the simulation of Alexandrov and Stenchikov for annual mean conditions. In some additional model studies, S.H. Schneider and S.L. Thompson (private communication, 1984) have found that the onset of this enhanced cross-equatorial circulation is sensitive to the reflectivity of the particulate cloud. For a sufficiently reflective cloud,* this qualitative change in the mean meridional circulation does not occur in the CCM.

The altered meridional circulation of Figure 7.11 would spread particulates rapidly upward and equatorward from the southern edge of the cloud. As the particulates spread, the heating would spread upward and equatorward with them. Since the particulate heating becomes more effective at the lower air densities of higher altitudes, the intensity of the thermally driven meridional circulation would increase as the particles rose. This positive feedback arising from the coupling between transport and heating has been observed in the GCM calculation of MacCraken and Walton (1984) and in two-dimensional model calculations (Haberle et al., 1983; M.C. MacCracken, private communication, 1984). In this simulation of a summer case in which an initial smoke cloud was located below 4 km, smoke from the equatorward edge of the cloud rapidly rose to 30 km, where its further evolution was influenced by the top of the model domain. Thus it is likely that this positive feedback mechanism could propel smoke to high levels, where its lifetime could be greatly lengthened. Even without the operation of the feedback mechanism, an enhanced meridional circulation like that produced by the CCM could transport smoke into the southern subtropics in 1 or 2 weeks. The large transport rates (up to 400 km/day) illustrated in Figure 7.11 developed without the feedback arising from transport of particulates, however.

Other Large-Scale Wind Systems

Large temperature differences between land and sea at low levels would produce effects that might be analogous to those occurring in midwinter at fairly high latitudes: prevalence of anticyclones with low-level outflow over the continents, surface cyclones over the high-latitude oceans, and development of frequent intense coastal storms, especially along eastern coasts of continents. At present, these effects must be regarded as speculative, although additional calculations with GCMs could narrow the uncertainties. Since many of the observed storms occurring in high-latitude coastal waters during winter and spring are quite small, model studies intended to simulate their behavior require higher horizontal spatial resolution than was used in the GCMs of either Covey et al. or Alexandrov and Stenchikov.

---

*Such a reflectivity could be produced by high-altitude dust injections and would itself imply a major long-term climate perturbation.

In addressing the question of interhemispheric transport, Covey et al. noted that the CCM results for spring showed strong localized cross-equatorial flows extending as far south as 30°S in the upper troposphere and lower stratosphere. Even in the unperturbed atmosphere, large-scale upper tropospheric troughs occasionally extend from mid-latitudes to the equator and beyond (Joung and Hitchman, 1982; Vincent, 1982; Huang and Vincent, 1983). These occurrences are more frequent during winter and over the oceans. Thus, if smoke reaches the upper troposphere, there is a good possibility that bands or streamers would be separated from the main cloud mass and stretched into the southern hemisphere even if the enhanced meridional circulation described above does not operate. The separation of bands of smoke from the southern edge of the main cloud would be associated with the complementary process: injection of streamers of clear air northward. In conjunction with spatial nonuniformity in the precipitation scavenging rate, this would ensure a degree of nonuniformity in optical depth even at long times after smoke injection, especially in the southern portion of the smoke-covered region.* More research is needed to understand transport near the southern edge of the clouds, both in the unperturbed and in the perturbed atmospheres.

Ultra-High Clouds

The intense heating in the upper portions of the particulate cloud should drive intense convection above the cloud. In the TTAPS model calculations, such convection does appear in the form of convective adjustment above the cloud. The temperature distribution shown in Figure 7.7 contains the effect of this convective adjustment; the temperature is shown to have increased in the region above the cloud as a consequence of upperward heat transport by convective adjustment.

Mixing in the convectively active layer would stir fine particles upward, thereby raising the altitude of maximum heating and further raising the altitude of the convective layer. This effect, which is in addition to any systematic tendency for large-scale circulation to raise the particulate cloud, was recognized by the TTAPS group, but not explicitly accounted for in their calculations.

Water vapor would also be mixed through the convective layer. At the top of the layer, a deep temperature minimum would develop, particularly as the convective elements are likely to overshoot the level of neutral buoyancy. The water vapor transported upward by the convection would be likely to condense to form a widespread cirriform cloud cover. The mass of material in this cloud would depend primarily on the water vapor concentrations at and above the base of the convective layer (near the temperature maximum). Such a cloud could

*In a series of unpublished GCM calculations, J. Mahlman (of Geophysical Fluid Dynamics Laboratory) has recently shown how these processes conspire to maintain nonuniformities in a simulated unperturbed atmosphere.

have significant radiative effects. Because ice crystals found in normal cirrus clouds tend to be of moderate size (with radii of several microns to a few tens of microns), and because ice is strongly absorbing in the infrared and reflective in the visible, normal cirrus generally has a larger influence on infrared radiation than on solar radiation. However, even a small increase in albedo due to such clouds would reduce the energy received by the atmosphere, so it is difficult to estimate the net climatic impact without detailed calculations.

As an example, suppose that water vapor from the base of the convective region is mixed upward uniformly through the convective layer with a mixing ratio of 100 ppmv, a representative value for air originating near the 200-mbar level. With adiabatic cooling of the rising air, condensation could begin near or slightly below the 50-mbar level. If the cloud extends 1 km above the condensation level and most of the water vapor in the cloud layer condenses, the resulting cloud mass would be about 7 g/m^2. The absorption cross section at 10-μm wavelength for spherical ice particles whose radii are a few microns or less is about 0.1 to 0.2 m^2/g (Bergstrom, 1973). Thus, in this example, an absorption optical depth of 1 for 10-μm radiation could develop for such a cloud. More work is needed to assess the significance of such ultra-high clouds. For example, if the absorbing particulate cloud moves upward, as a result of self-induced circulation or mixing, the infrared opacity of such an elevated cirrus layer would be correspondingly smaller.

Longer Term Effects on Climate

If nuclear war injections of smoke were as large as those of the NRC baseline case, longer term meteorological effects, extending beyond the time at which most of the smoke is removed from the atmosphere, might occur. Such effects could arise from changes in the distribution of snow, sea ice, and vegetation cover, which would cause changes in surface albedo, thermal inertia, and evapotranspiration potential. It is also possible that persistent changes in ocean current systems leading to changes in sea surface temperature distributions would be produced. The upward mixing of water vapor by convection to altitudes above 10 km could also have significant long-term climatic implications. Such possibilities are extremely difficult to evaluate, particularly because shorter term effects themselves are highly uncertain. However, Robock (1984) has recently attempted to assess some of these effects using an EBCM with snow and ice albedo feedback and sea ice thermal inertia and meltwater feedbacks included in the model (Robock, 1983). Applying this model to the TTAPS scenario, he found depressed surface temperatures persisting but gradually ameliorating over several years in northern, middle, and high latitudes, primarily as a result of an increase in the surface covered by sea ice with a corresponding reduction in thermal inertia of the northern high-latitude oceans.

An effect that could be significant but would favor warming of high-latitude surface temperatures is the depression of snow and ice albedo due to the fallout of smoke particles. If as little as 10 to 20 Tg of smoke particles was to fall out over the Arctic during the course of a few months and if the smoke particles were mixed with no more than the normal amount of snowfall, they could have a very significant effect on snow albedo (Warren and Wiscombe, 1984). The actual importance of this effect is difficult to evaluate, however, since it depends on many detailed processes, such as the exact timing of smoke and snow fallout events, washout of smoke particles due to surface melting on snow or ice, and changes in the morphology of the snow or ice surfaces.

Such longer term effects are difficult to investigate, but they should not be ignored.

ANALOGS

Of necessity the previous discussion relies heavily on model results, supplemented by occasional references to our understanding of how the undisturbed atmosphere behaves. Confidence in these results can be enhanced by examining natural situations where some of the key processes and their effects can be seen. Indeed, bare model results in the absence of such natural analog situations would be quite unconvincing to many observers. In this section several such natural analogs are examined.

Arctic Haze

Recent research has shown that there is a remarkable amount of aerosol pollution in the central Arctic, especially during spring (Patterson et al., 1982; Rosen and Novakov, 1983). A major component of this pollution is a fine particle mode (particle mode diameter of about 0.4 μm), which in turn is rich in soot carbon. This material has been detected near the surface and in layers at elevations as high as 5 km (Hansen and Rosen, 1984; Radke et al., 1984). The particles in such elevated layers, following essentially quasi-isentropic trajectories,* must have originated at distant mid-latitude pollution sources, and they must in some cases have been in transit for many days. Thus the properties of these particles provide valuable information on the aging of carbonaceous particulates in the unperturbed atmosphere. Microscopic analysis and analysis of the optical properties of these particles indicate that the soot particles sometimes occur internally mixed in a nonabsorbing material, probably sulfate (A.D. Clarke, private communication, 1984). The polluted layers also contain nonabsorbing

*Heating can probably be neglected to first order in considering the transport of these particles, so that they would tend to move approximately on surfaces of constant specific entropy. Since these slope upward toward the pole, pollutants originating near the surface can reach the middle troposphere in the Arctic.

particles unmixed with carbonaceous material so that the mean single scattering albedo of all particles varies around 0.86 (Clarke et al., 1984). This value is considerably higher than that of the postulated nuclear war smoke clouds, though nevertheless the polluted layers are quite strongly absorbing. In relating these aerosols to the smoke that could be produced by burning cities, it is important to keep in mind that the former are probably produced in pollution plumes that are rich in sulfur and not particularly black at the source; the smoke from burning cities is likely to be much blacker initially and throughout its life in the atmosphere.

Elemental carbon several days removed from its sources has also been found to be an important component of the fine particle mode in the marine boundary layer over the Atlantic (Andreas, 1983). Although highly variable, typical soot fractions of the fine particle mass were about 40 percent. Further experimental studies of the fine particle mode in regions remote from pollution sources should provide valuable information on the mechanisms, rates, and consequences of the aging of carbonaceous particles in the undisturbed atmosphere. This information is a necessary prerequisite to understanding the implications of soot aging for the consequences of nuclear war.

Plumes from Large Forest Fires

There are a number of accounts of observations of forest fire plumes at large distances from their sources (see Chapter 5). Lyman (1918), for example, documents a case in which smoke from large fires in Minnesota darkened the sky over much of the northeastern United States and southeastern Canada. Shostakovitch (1925) gives a dramatic account of the obscuration persisting for more than a month due to the Siberian forest fires of 1915.

Wexler (1950) provides a well-documented account of the plume from a large number of forest fires burning within a 40,000 km^2 area of northwest Alberta and northeast British Columbia (although the extent of the area that actually burned is unclear from Wexler's account). Wexler describes events during the period September 24 to 30, 1950. Within 2 days of the beginning of the most intense phase of burning, the plume had reached Washington, D.C. Within 5 days, it had been observed over all of Canada except the far northeast and far west, over almost the entire United States east of the Mississippi River plus Minnesota and the Dakotas, and had stretched across the North Atlantic and had been observed throughout Western Europe from Portugal to Norway (Figure 7.12).

At Washington, D.C., the smoke occurred in a layer between the 2.5 and 5 km altitudes bounded above and below by inversions, and was estimated by Wexler to have reduced the total incident solar radiation by as much as 54 percent. Associated with this reduction was a decrease in maximum temperature that Wexler estimated to be an average of 4°C for 4 days. Smith (1950) quotes an estimate by Fritz that the maximum temperature was reduced by as much as 6°C, with no compensating rise in minimum temperature. By the time the plume had reached England, it

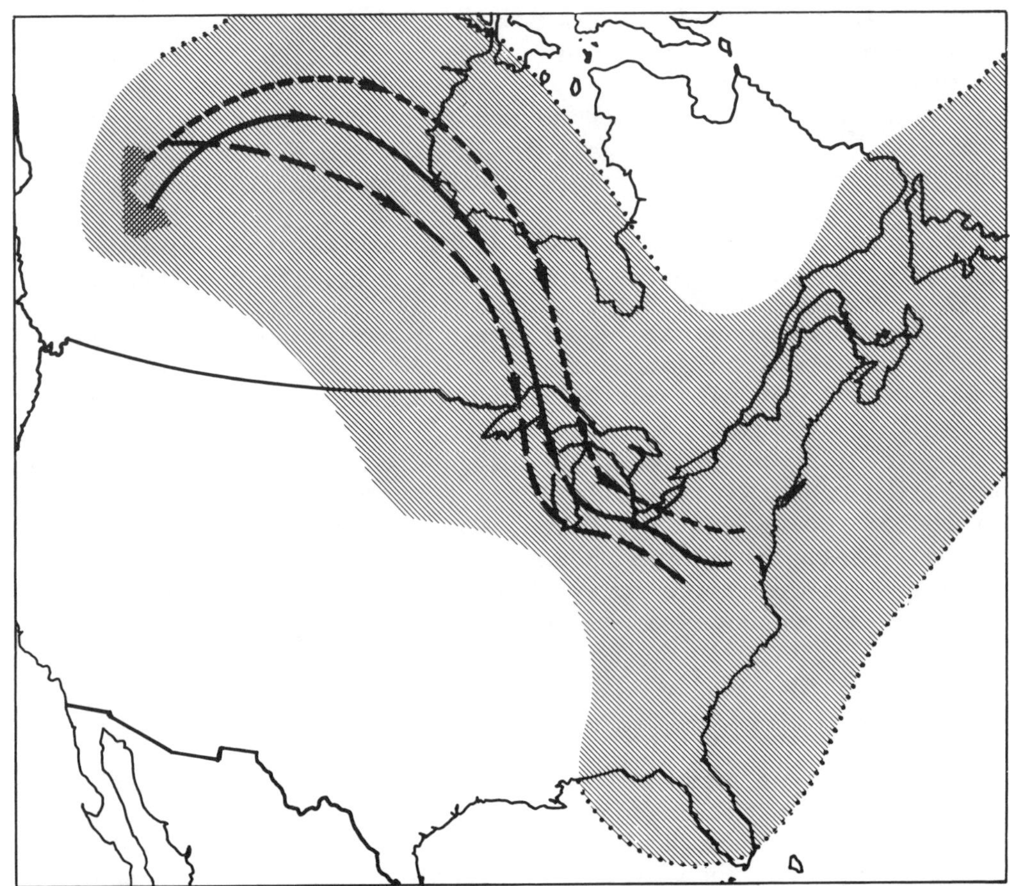

FIGURE 7.12 The hatched area represents the region over which smoke was observed from the western Canada forest fires of September 1950 (exclusive of observations from Western Europe). The boundary of this area is dotted where it is tentative. The darkened areas in western Canada are the areas in which the fires occurred, and the curves mark calculated trajectories for smoke reaching the vicinity of Washington, D.C., by September 24, two days after the most intense burning episode. (From Smith, 1950.)

appears to have risen to an altitude range of 10 to 12 km (Bull, 1951).

These incidents illustrate the rapid spread of fire plumes from relatively small areas. They also show that such plumes can have dramatic optical effects and can influence surface temperatures thousands of kilometers from the source. Such forest fire plumes are not necessarily highly absorbing for solar radiation, however. The reduction in solar radiation and the surface temperature decreases observed at Washington were probably due largely to reflection rather than absorption of sunlight by the cloud. As discussed in Chapter 5, urban fires are likely to produce much blacker smoke, and to produce much larger optical depths and reductions in solar radiation at the surface.

Early Plume from the Mount St. Helens Eruption

The paroxysmal eruption of Mount St. Helens on May 18, 1980, produced a large plume of ash that spread rapidly across eastern Washington and into Idaho and Montana during the day following the eruption. Rapid daytime temperature decreases were observed beneath the plume. By comparing observed and forecast temperatures under the plume with those in the surroundings, Mass and Robock (1982) argued that the plume produced a drop in the maximum temperature of up to 8°C. However, during the following night, as the plume drifted over Montana, increases in minimum temperature of about the same magnitude were observed. Evidently, the substantial reduction in solar radiation produced by the plume was compensated by a corresponding increase in the downward infrared radiation.

The properties of the ash particles in this early volcanic plume were quite different from those of the smoke particles of the nuclear war scenarios. The Mount St. Helens ash particles had high single scattering albedos, and the size distributions had maximum diameters between 1 and 10 μm. The plume is estimated to have contained about 2 Tg of ash particles with diameters greater than 2 μm, but less than 10^{-2} Tg of particles with diameters less than 2 μm (Hobbs et al., 1982), so it is not surprising that the plume was an effective emitter of infrared radiation at this stage of its evolution.

These observations illustrate the rapidity with which such plumes can influence surface temperatures, and they serve to focus attention on the role of the ratio of infrared to visible absorptivity of particles in the nuclear war scenarios.

Sahara Dust Plumes, the "Harmattan"

Sahara dust carried over West Africa and the tropical Atlantic Ocean by northeasterly and easterly winds provides another natural analog for some facets of the nuclear war problem. Outbreaks of dust over the Atlantic can produce extinction optical depths of about 1 over areas of 10^6 km^2 (Carlson and Caverly, 1977; Carlson and Benjamin, 1980). As much as 8 Tg of dust may be involved in a large outbreak (Carlson, 1979), and strong heating occurs in the dusty layer. Suppression of convection has been noted when Sahara dust in the middle troposphere is transported over the tropical Atlantic.

During the dry season in West Africa, the prevailing northeasterly wind, which is often laden with dust, is known as the "harmattan." Brinkman and McGregor (1983) report harmattan events in Nigeria with optical depths up to 2 and associated reductions in daily mean total solar radiation of 28 percent. They also report temperature decreases of up to 6°C for these events, although this is representative of the depression of the maximum rather than the daily mean temperature.

Although these dust particles are probably generally much larger than the stratospheric dust particles and are both larger and more reflective than the smoke particles of the nuclear war scenarios, these observations show that such aerosols do have a rapid effect on surface temperatures. They also show that such particles, even though less absorbing than smoke, produce elevated heated layers that can act to suppress convection.

Martian Global Dust Storms

It is now known that the planet Mars is subject to occasional global-scale dust storms in which dust spreads over most of the planet with mean optical depths of order 5. Martian dust is somewhat more absorbing at visible wavelengths than typical terrestrial dusts, so that the absorptivity for these situations is intermediate between values for nuclear war scenarios with dust only and those with both smoke and dust. Consequently, the scale of the associated optical perturbation is within the range of interest. These events produce temperature increases in the upper part of the dusty layer of order 80°C over much of the planet. Temperature decreases at both subtropical and mid-latitude sites have also been observed in connection with these events (Martin and Kieffer, 1979; Pollack et al., 1979; Ryan and Henry, 1979). The vertical profile of temperature changes associated with these events resembles that of the nuclear war scenarios except that the decrease in surface temperature is less. This is partly because Martian dust is much less absorbing in the visible than smoke, but, probably more important, it is because the "greenhouse" effect is at most very weak on Mars, so that the "antigreenhouse" effect at the surface due to the absorbing cloud is not very pronounced (see page 149).

These dust storms do not occur every Martian year. When they do occur, it is during southern hemisphere summer, Mars perihelion season, when dust generated locally in the summer subtropics is swept upward to great heights in the rising branch of the mean meridional circulation and then is swept rapidly poleward, reaching high latitudes of the opposite hemisphere within a few days (Haberle et al., 1982). Proper phasing between dust injection and the meridional circulation is an essential feature of this phenomenon; dust injected into the normally subsiding branch of the tropical mean meridional circulation remains close to the latitude of injection.

The analogy to the nuclear war scenarios should not be pressed too far. The total amount of material involved in the Martian dust storms is larger (Toon et al., 1977), but the particle sizes are larger so they are less efficient optically; precipitation processes are not active on Mars; and the global dust storms are driven by heating per unit mass of atmosphere that is larger than the largest reasonable values for the nuclear war smoke clouds. Nevertheless, Mars does provide a natural example of the thermal structure of an "antigreenhouse" atmosphere and of rapid meridional spread of particulates by an enhanced thermally driven meridional circulation.

SUMMARY

None of the natural situations described above bears a close resemblance to the atmospheric condition that is likely to prevail following a full-scale nuclear war. Nevertheless, each has elements that tend to support various conclusions drawn from the models.

In sum then, the various model results in concert with a limited set of observations of related natural phenomena provide a basis for concluding that a nuclear war scenario like the NRC baseline case could produce large temperature decreases near the surface and temperature increases aloft for a period of weeks to months following the event (cf. the two- and three-dimensional model results summarized in Tables 7.3 and 7.4). Moreover, rapid spreading of particulates into the tropics and even into the southern hemisphere is a real possibility. These conclusions are contingent upon the assumptions that a substantial fraction of the smoke particles produced by burning cities would survive early scavenging and coagulation, and that subsequent aging and scavenging processes would not remove submicron smoke particles distributed throughout the middle and upper troposphere at a removal rate* greater than about $(2 \text{ weeks})^{-1}$. Because of optical saturation due to the high absorptivity of smoke, the climatic effects are likely to be insensitive to moderate changes in smoke or absorptivity about the baseline values. However, lower values of either of these quantities by a factor of about 4 would lie near the edge of the saturation regime, and climatic effects would decrease rapidly for large reductions. Climatic effects are also sensitive to the removal rate of smoke. If middle and upper tropospheric rates were as large as $(1 \text{ week})^{-1}$ temperature perturbations would be considerably moderated although still significant (see the "fast rainout" used in Figure 7.6). Improvements in the models are needed, particularly to investigate further the effects of realistic transport and dispersion of smoke and dust in the perturbed atmosphere, the infrared opacity of the smoke, diurnal and seasonal effects, and the possible roles of ground fog and stratus and of ultra-high clouds forming at the top of the convective layer that may be driven by absorption of solar radiation in smoke and dust clouds. Long-term effects arising from possible changes in the properties of the underlying surface also require further study.

REFERENCES

Ackerman, T.P., and O.B. Toon (1981) Absorption of visible radiation in atmosphere containing mixtures of absorbing and non-absorbing particles. Appl. Opt. 20:3661-3668.

Aleksandrov, V.V., and G.L. Stenchikov (1983) On the modeling of the climatic consequences of the nuclear war. _In_ Proceedings on Applied Mathematics. Moscow: Computing Center of the Academy of Sciences USSR.

---

*Removal rate is defined in Table 7.2.

Ambio (1982) Nuclear war: The aftermath. 11(2/3):75-176.
Andreas, M.O. (1983) Soot carbon and excess fine potassium: Long-range transport of combustion-derived aerosols. Science 220:1148-1151.
Bergstrom, R.W. (1973) Extinction and absorption of atmospheric aerosol as a function of particle size. Contrib. Atmos. Phys. 46:223-234.
Brinkman, A.W., and J. McGregor (1983) Solar radiation in dense Saharan aerosol in northern Nigeria. Quart. J. Roy. Meteorol. Soc. 109:831-847.
Bull, G.A. (1951) Blue sun and moon. Meteorol. Mag. 80:1-4.
Carlson, T.N. (1979) Atmospheric turbidity in Saharan dust outbreaks as determined by analysis of satellite brightness data. Mon. Weather Rev. 107:322-335.
Carlson, T.N., and S.G. Benjamin (1980) Radiative heating rates for Saharan dust. J. Atmos. Sci. 37:193-213.
Carlson, T.N., and R.S. Caverly (1977) Radiative characteristics of Saharan dust at solar wavelengths. J. Geophys. Res. 82:3141-3152.
Cess, R.D. (1984) Nuclear war: Illustrative effects of atmospheric smoke and dust upon solar radiation. Unpublished manuscript. Laboratory for Planetary Atmosphere Research, State University of New York, Stony Brook.
Cess, R.P., G.L. Potter, and W.L. Gates (1984) Climatic impact of a nuclear exchange: Sensitivity studies using a general circulation model. Paper presented at the 4th Session of the International Seminar on Nuclear War, Erice, Sicily. Aug. 19-24, 1984.
Chen, C.-S., and H.D. Orville (1977) The effects of carbon black dust on cumulus scale convection. J. Appl. Meteorol. 16:401-412.
Clarke, A.D., R.V. Charlson, and L.F. Radke (1984) Airborne observations of Arctic aerosol 4: Optical properties of Arctic haze. Geophys. Res. Lett. 11:405-408.
Covey, C., S.H. Schneider, and S.L. Thompson (1984) Global atmospheric effects of massive smoke injections from a nuclear war: Results from general circulation model simulations. Nature 308:21-31.
Crutzen, P., I.E. Galbally, and C. Brühl (1984) Atmospheric effects from post-nuclear fires. Climatic Change (in press).
Gates, W.L., and M.E. Schlesinger (1977) Numerical simulation of the January and July global climate with a two-level atmospheric model. J. Atmos. Sci. 34:36-76.
Golitsyn, G.S., and A.S. Ginsburg (1984) Comparative estimates of the climatic consequences of Martian dust storms and possible nuclear war. Paper presented at the Conference, The World After Nuclear War, Oct. 31 to Nov. 1, 1983. Institute of Atmospheric Physics of the Academy of Sciences USSR.
Haberle, R.M., C.B. Leovy, and J.B. Pollack (1982) Some effects of global dust storms on the atmospheric circulation of Mars. Icarus 50:322-367.
Haberle, R.M., T.P. Ackerman, and O.B. Toon (1983) The dispersion of atmospheric dust and smoke following a large-scale nuclear exchange. Paper presented at the Fall 1983 Meeting of the American Geophysical Union, San Francisco, G.T. Wolff and R.L. Klimisch, eds. New York: Plenum. Pages 379-391.

Hamill, P., R.P. Turco, C.S. Kiang, O.B. Toon, and R.C. Whitten (1982) On the formation of sulfate aerosol particles in the stratosphere. J. Aerosol Sci. 13:565-581.

Hansen, A.D.A., and H. Rosen (1984) Vertical distributions of particulate carbon, sulfur, and bromine in Arctic haze and comparison with Barrow, Alaska. Geophys. Res. Lett. 11:381-384.

Held, I., and A. Hou (1980) Non-linear axially symmetric circulations in a nearly inviscid atmosphere. J. Atmos. Sci. 37:515-533.

Hobbs, P.V., J.P. Tuell, D.A. Hegg, L.F. Radke, and M.W. Eltgroth (1982) Particles and gases in the emissions from the 1980-1981 volcanic eruptions of Mt. St. Helens. J. Geophys. Res. 87:11062-11086.

Holton, J.R. (1975) The dynamic meteorology of the stratosphere and mesosphere. Meteorological Monographs, No. 34. Boston: American Meteorological Society.

Holton, J.R. (1979) An Introduction to Dynamic Meteorology. 2nd ed. New York: Academic. 391 pp.

Huang, H.-J., and D.G. Vincent (1983) Major changes in the circulation features over the South Pacific during FGGE, 10-27 January 1979. Mon. Weather Rev. 111:1611-1618.

Joung, C.-H., and M.H. Hitchman (1982) On the role of successive downstream development in East Asian polar air outbreaks. Mon. Weather Rev. 110:1224-1237.

Lee, K.T. (1983) Generation of soot particles and studies of factors controlling soot light absorption. Ph.D. thesis, Department of Civil Engineering, University of Washington, Seattle.

Levy, H., II, J.D. Mahlman, and W.J. Moxim (1980) Three-dimensional tracer structure and behaviour as simulated in two ozone precursor experiments. J. Atmos. Sci. 37:655-685.

Liou, K.-N. (1980) An Introduction to Atmospheric Radiation. New York: Academic. 392 pp.

Lyman, H. (1918) Smoke from the Minnesota forest fires. Mon. Weather Rev. 46:506-509.

MacCracken, M.C. (1983) Nuclear war: Preliminary estimates of the climatic effects of a nuclear exchange. Paper presented at the International Seminar on Nuclear War, 3rd Session: The Technical Basis for Peace, Ettore Majorana Centre for Scientific Culture, Erice, Sicily, Aug. 12-23, 1983.

MacCracken, M.C., and J. Walton (1984) The effects of interactive transport and scavenging of smoke on the calculated temperature change resulting from large amounts of smoke. Paper presented at the International Seminar on Nuclear War, 4th Session, Erice, Sicily, Aug. 19-24, 1984.

Mahlman, J.D., and W.J. Moxim (1978) Tracer simulation using a global general circulation model: Results from a mid-latitude instantaneous source experiment. J. Atmos. Sci. 35:1340-1374.

Martin, T.Z., and H. Kieffer (1979) Thermal infrared properties of the Martian atmosphere. 2. The 15-μm band measurements. J. Geophys. Res. 84:2843-2852.

Mass, C., and A. Robock (1982) The short-term influence of the Mount St. Helens volcanic eruption on surface temperature in the northwest United States. Mon. Weather Rev. 110:614-622.

Massie, S.T., and D.M. Hunten (1981) Stratospheric eddy diffusion coefficients from tracer data. J. Geophys. Res. 86:9859-9868.

Ogren, J.A. (1982) Deposition of particulate elemental carbon from the atmosphere. In Particulate Carton: Atmospheric Life Cycle, edited by G.T. Wolff and R.L. Klimisch. New York: Plenum.

Ogren, J.A., and R.J. Charlson (1983) Elemental carbon in the atmosphere: Cycle and lifetime. Tellus 358:241-254.

Patterson, E.M., B.T. Marshall, and K.A. Rahn (1982) Radiative properties of the Arctic aerosol. Atmos. Environ. 16:2967-2977.

Pollack, J.B., O.B. Toon, and B.N. Khare (1973) Optical properties of some terrestrial rocks and glasses. Icarus 19:372-389.

Pollack, J.B., O.B. Toon, C. Sagan, A. Summers, B. Baldwin, and W. van Camp (1976) Volcanic explosions and climate change: A theoretical assessment. J. Geophys. Res. 81:1071-1083.

Pollack, J.B., D.S. Colburn, F.M. Flasar, R. Kahn, C.E. Carlston, and D. Pidek (1979) Properties and effects of dust particles suspended in the Martian atmosphere. J. Geophys. Res. 84:2929-2945.

Pollack, J.B., O.B. Toon, T.P. Ackerman, C.P. McKay, and R.P. Turco. (1983) Environmental effects of an impact-generated dust cloud: Implications for the Cretaceous-Tertiary extinctions. Science 219:287-289.

Radke, L.F., J. Lyons, D. Hegg, P.V. Hobbs, and I. Bailey (1984) Airborne observations of Arctic aerosols. 1. Characteristics of Arctic haze. Geophys. Res. Lett. 11:393-396.

Ramaswamy, V., and J. Kiehl (1984) Sensitivity of the radiative forcing due to large loadings of smoke and dust aerosols. Manuscript, National Center for Atmospheric Research, Boulder, Colo. (Submitted to J. Geophys. Res.)

Robock, A. (1983) Ice and snow feedbacks and the latitudinal and seasonal distribution of climate sensitivity. J. Atmos. Sci. 40:986-997.

Robock, A. (1984) Snow and ice feedbacks for prolonged effects of nuclear winter. Nature 310:667-670.

Rosen, H., and T. Novakov (1983) Combustion-generated carbon particles in the Arctic atmosphere. Nature 306:768-778.

Ryan, J.A., and R.M. Henry (1979) Mars atmospheric phenomena during major dust storms as measured at the surface. J. Geophys. Res. 84:2821-2829.

Schneider, E.K. (1983) Martian great dust storms: Interpretive axially symmetric models. Icarus 55:302-331.

Sellers, W.D. (1973) A new global climate model. J. Appl. Meteorol. 12:241-254.

Shostakovitch, V.B. (1925) Forest conflagrations in Siberia. J. Forestry 23:365-371.

Smith, C.D., Jr. (1950) The widespread smoke layer from the Canadian forest fires during late September 1950. Mon. Weather Rev. 78:180-184.

Thompson, S.L., V.V. Aleksandrov, G.L. Stenchikov, S.H. Schneider, C. Covey, and R.M. Chervin (1984) Global climatic consequences of nuclear war: Simulations with three-dimensional models. Ambio 13 (in press).

Toon, O.B., and T.P. Ackerman (1981) Algorithms for the calculation of scattering by stratified spheres. Appl. Opt. 20:3657-3660.

Toon, O.B., J.B. Pollack, and C. Sagan (1977) Physical properties of the particles comprising the Martian dust storm of 1971-1972. Icarus 30:663-696.

Toon, O.B., R.P. Turco, P. Hamill, C.S. King, and R.C. Whitten (1979) A one-dimensional model describing aerosol formation and evolution in the stratosphere: II. Sensitivity studies and comparison with observations. J. Atmos. Sci. 36:718-736.

Turco, R.P., P. Hamill, O.B. Toon, R.C. Whitten, and C.S. Kiang (1979) A one-dimensional model describing aerosol formation and evolution in the stratosphere. I. Physical properties and mathematical analogs. J. Atmos. Sci. 36:699-717.

Turco, R.P., O.B. Toon, R.C. Whitten, J.B. Pollack, and P. Noerdlinger (1982) An analysis of the physical, chemical, optical and historical impacts of the 1908 Tunguska meteor fall. Icarus 50:1-52.

Turco, R.P., O.B. Toon, T. Ackerman, J.B. Pollack, and C. Sagan (1983a) Nuclear winter: Global consequences of multiple nuclear explosions. Science 222:1283-1293.

Turco, R.P., O.B. Toon, T. Ackerman, J.B. Pollack, and C. Sagan (1983b) Global Atmospheric Consequences of Nuclear War. Interim Report. Marina del Rey, Calif.: R&D Associates. 144 pp.

Vincent, D.G. (1982) Circulation features over the South Pacific during 10-18 January 1979. Mon. Weather Rev. 110:981-993.

Walter, B.A. (1980) Wintertime observations of roll clouds over the Bering Sea. Mon. Weather Rev. 108:2024-2031.

Walton, J.J., and M.C. MacCracken (1984) Preliminary report on the global transport model Grantour. Unpublished report. Lawrence Livermore National Laboratory, Livermore, Calif.

Warren, S., and W. Wiscombe (1984) Dirty snow after nuclear war. Nature (in press).

Washington, W.M., ed. (1982) Documentation for the Community Climate Model (CCM) Version O. Boulder, Colo.: National Center for Atmospheric Research. (NTIS PB82-194192.)

Wexler, H. (1950) The great smoke pall--September 24-30, 1950. Weatherwise (Dec.):129-142.

8
Use of Climatic Effects of Volcanic Eruptions and Extraterrestrial Impacts on the Earth as Analogs

Very large explosive volcanic eruptions and asteroid or meteor impacts can inject large amounts of dust high into the atmosphere. It is important therefore to assess the extent to which data provided by such events can be useful in the attempt to understand the atmospheric modification that would follow a nuclear exchange. The committee did not, however, find any unambiguous evidence provided by volcanic and impact events to support or refute a conclusion that nuclear war may seriously affect the world's climate. No recent natural events have been energetic enough to provide more than a small atmospheric perturbation; furthermore, the only investigations of earlier, larger events, whose goals included dust lofting estimates have been those associated with the hypothesis that a very large meteor caused the extinction at the Cretaceous-Tertiary boundary some 65 million years ago. That event would have been more energetic than the baseline exchange by a factor of more than 10^4. Nevertheless, an account is included here of those aspects of volcanic and natural impact events that, if there were more data available, would be pertinent.

It is important to consider the similarities and differences between volcanic explosions and nuclear explosions, and between extraterrestrial impacts and nuclear explosions. One must keep in mind that substantial unknowns exist in our understanding of the effects of these events on the terrestrial ecosystem, which parallel in many ways the uncertainties in our understanding of the potential effects of nuclear war. For example, it is only for the past 100 years that a reliable record exists of the optical depths of clouds of volcanic origin.

VOLCANIC ERUPTIONS

Since large volcanic eruptions could introduce quantities of material into the atmosphere comparable to those from a nuclear war, it is pertinent to ask if we can use evidence from volcanic explosions to empirically determine the climatic effects that would be caused by a nuclear war. It turns out that volcanoes do not prove very useful in this regard.

The largest explosive volcanic eruptions in the last 200 years have occurred at Tambora, Indonesia, in 1815; at Krakatau, Indonesia, in 1883; and at Katmai, Alaska, in 1912 (Simkin, 1981). Each of these eruptions produced 10^{16} to 10^{17}g of 700° to 900°C volcanic fragments within a few hours to a few days. All three eruptions produced large ash clouds that reached into the stratosphere. The total thermal energies of these eruptions amounted to 10^{19} to 10^{20} joules (J), of which only about 1 percent was converted to mechanical energy (as steam blasts, the initial velocities of ejected material, and the buoyant lift of the ash cloud). This volcanic mechanical energy (10^{17} to 10^{18} J) is equivalent to 25 to 250 Mt of nuclear energy. However, the release of volcanic energy takes place over minutes to days, and hence the power (energy release per second) of historic volcanic eruptions has been much less than would be the power of a single-megaton nuclear explosion. Nevertheless, some volcanic blasts devastate areas similar in size to areas that would be affected by nuclear explosions, and they loft large amounts of dust and gases into the stratosphere.

Prehistoric volcanic eruptions have produced thick blankets of explosive volcanic debris called ash flows covering areas of up to 10,000 km^2 with masses of up to 2 x 10^{18} g. Most of the enormous thermal energy in these huge eruptions (2 x 10^{21} J, equivalent to 500,000 Mt) is retained in the hot fragmental debris blanket and is slowly dissipated over tens to hundreds of years.

In addition to energy release patterns, another important distinction between volcanic explosion clouds and nuclear explosion clouds involves composition. Volcanic clouds contain large quantities of sulfur gases. However, both volcanic explosion clouds and near-surface nuclear explosion clouds contain silicate glass and silicate mineral particles.

Very large volcanic eruptions occur infrequently, on the average of about once every 10,000 to 100,000 years. They occur as locally isolated events, whereas the 25,000 potential nuclear explosions assumed in this report occur nearly simultaneously over large areas.

Historic volcanic explosions have not generated large forest or brush fires, and the thick ash flow blankets of very large prehistoric eruptions would have tended to smother fires. Hence the "soot" problem that results from multiple nuclear explosions probably has no counterpart in volcanic eruptions.

The effects of gases and dust from volcanic eruptions on climate have been a subject of speculation ever since Benjamin Franklin alleged that the Laki eruption in Iceland in 1783 had caused a "dry fog" in Europe with attendant cold weather and poor crops (Humphreys, 1940). Some investigators have concluded that increased volcanism was a major cause of the Pleistocene ice age (Kennett and Thunnel, 1977); others have argued that apparent worldwide average temperature drops of up to 1°C in the year following the large explosive eruptions of Tambora Volcano (Indonesia, 1815) and Krakatau Volcano (Indonesia, 1883) either could be errors of measurement (or analysis), or if real, could be coincidental to normal fluctuations in average world temperature (Landsberg and Albert, 1974).

TABLE 8.1 Estimates of Volcanic Products in the Stratosphere Following Recent Eruptions (data as mass in grams, size of middle two-thirds of particles, l/e folding times)

Eruption, Date (volume, dense rock equivalent, height of column)	Silicates		Nonsilicate	
	All	<2 μm only	(Sulfate)	S Gases
Agung, 1963[16] (0.9 km^3, 23 km)		1×10^{13} g[1] 2-0.2 μm[2]	2×10^{13} g[1] 0.2-0.45 μm[2] ~1 year	0.6×10^{13} g[1]
Fuego, 1984[17] (0.1 km^3, >1.4 km)	<1.4×10^{13} g[3] 30-1.0 μm days	7×10^9 g[3,16] (2-0.2 μm)[3]	3-6 $\times 10^{12}$ g[1,4] <1.0 μm 11.6 months[11]	1.6×10^{12} g[3]
Mount St. Helens, 1980[12] (0.3 km^3, >24 km)	<1.5×10^{14} g[12] 30-1.0 μm[5,6] days[7]	<1-2 months[6]	2.5-5 $\times 10^{11}$ g[6] 0.2-0.6 μm[7] ?[11]	~2×10^{12} g[6]

El Chichon, 1982[13,19] ($0.38\ km^3$, >17 km)	2.2×10^8 to 2×10^{14} g[13,18,20] 30-1.0 μm[14] weeks or days[14]		2×10^{13} g[8] 0.1-0.7 μm[8,10] ?[15]	3.8-13.4 $\times 10^{12}$ g[9,19]

NOTES:

1. Cadle et al. (1976)
2. Mossop (1963, 1965)
3. Murrow et al. (1980)
4. Lazrus et al. (1979)
5. Rose and Hoffman (1982)
6. Newell and Deepak (1982), M. Milan (personal communication)
7. Chuan et al. (1981), Farlow et al. (1981)
8. Hofmann and Rosen (1983)
9. Krueger (1983)
10. Hirono and Shibata (1983)
11. Sedlacek et al. (1983)
12. Rose et al. (1983b)
13. Varekamp et al. (1984)
14. Clanton et al. (1982), Chuan et al. (1984)
15. Iwasaki et al. (1983)
16. Self et al. (1981)
17. Rose et al. (1978)
18. Sigurdsson et al. (in press)
19. Evans and Kerr (1983)
20. Gooding et al. (1983)

SOURCE: Compiled by W.I. Rose.

Most present investigators of the influence of volcanic eruptions on climate (Pollack et al., 1976; Robock, 1978; Toon and Pollack, 1982) agree that there is a measurable effect and that it is largely caused by the sulfuric acid aerosol particles that form in the stratosphere from sulfur dioxide gases in volcanic ash clouds that reach stratospheric altitudes.

Actual sampling of the amount of dust and gases from volcanic eruptions reaching the stratosphere began following the eruption of Agung Volcano (Indonesia, 1963). Data on silicate dust (mainly volcanic glass and small silicate mineral fragments) and sulfur gases reaching the stratosphere from recent eruptions of Agung, Fuego (Guatemala), Mount St. Helens, and El Chichon (Mexico) volcanoes are listed in Table 8.1. Rough estimates of the amounts of dust and sulfur gases reaching the stratosphere from the eruptions of Tambora, Krakatau, and Agung are given in Table 8.2. Since the Tambora and Krakatau eruptions produced 10 to 100 times more volcanic debris than the Agung, Fuego, Mount St. Helens, and El Chichon eruptions, it is apparent from the tables that larger volcanic eruptions do not generate linearly proportional amounts of dust and gases that reach the stratosphere. One probable reason for this lack of proportionality is that larger eruptions produce denser ash clouds, which are less buoyant. Both Settle (1978) and Wilson et al. (1978) have demonstrated by theory and observation of historic volcanic eruptions that the height of explosive eruption clouds increases as the rate of emission of fragmental volcanic material increases. However, Wilson et al. calculate that eruption rates of about 10^6 m^3/s will generate maximum ash cloud heights of 55 km. Larger eruption rates will produce dense clouds that will fall back under their own weight before reaching this maximum altitude.

This lack of linear scaling is important in considering the possible climatic effects of extremely large volcanic explosions in prehistoric times. In the last 2 million years, there have been six explosive volcanic eruptions in the western United States (three at Yellowstone; two at Valles Caldera, New Mexico; and one at Long Valley, California) that have produced 100 to 2000 km^3 of fragmental volcanic material (ash flows) during apparent time intervals of a few hours to weeks (Francis, 1983). An estimate for the total of both fine dust and sulfate aerosols injected into the stratosphere by a single very large volcanic eruption is 10^{15} g; however, there is a high degree of uncertainty in this estimate. Linearly proportional scaling of the 10^{13} g of stratospheric dust and aerosols produced by the El Chichon eruption up to a Yellowstone-type eruption yields an estimate of 10^{16} g as an upper limit. The approximately 10^{14} g fallout of sulfate from the Tambora eruption (Table 8.2) provides an apparent lower limit. However, large sulfate loadings may be self-limiting due to the nonlinear dependence of growth rate and sedimentation of injected sulfur. Therefore it is not clear that volcanoes have ever exceeded the apparent lower limit. Loading the global stratosphere with 10^{15} g of fine dust and sulfate aerosols from a great volcanic explosion would produce a worldwide average temperature drop of about 10°C for

TABLE 8.2 Comparison of Estimates of Fine Dust, Aerosols, and Sulfate Fallout

Eruption, Year, and Volume[a]	Total <2-μm Dust[b] (g)	Total Aerosol[c] (g)	Global SO_4 Fallout[d] (g)
Tambora, 1815 About 150 km^3	1.5×10^{13} to 1.2×10^{18} (150)	? (?)	1.5×10^{14} (7.5)
Krakatau, 1883 About 20 km^3	2×10^{13} to 1.6×10^{14} (20)	3×10^{13} (~3)	5.5×10^{13} (~3)
Agung, 1963 About 1 km^3	1×10^{11} to 8×10^{13} (1)	0.9×10^{13} (1)	2×10^{13} (1)

[a]Volumes of eruptions are expressed as bulk volumes of near-source and distal ejecta.
[b]Using the method of Murrow et al. (1980), assumes an average bulk density of fine ash of 1 g/cm^3. This volume represents the total mass of <2-μm-diameter dust ejected into the atmosphere. Only a fraction of this dust entered the stratosphere. Evidence from Mount St. Helens eruptions (Rose and Hoffman, 1982) demonstrates that a large, but as yet undetermined, portion of the fine dust will be quickly removed by particle aggregate formation.
[c]Deirmendjian (1973). Refers to stratospheric loading.
[d]Hammer et al. (1980). Refers to stratospheric loading.

NOTE: Figures in parentheses represent relative quantities of dust, aerosol, and sulfate fallout, with Agung (1963) as base figure.

SOURCE: Rampino and Self (1982).

several months. This hypothetical climate anomaly is calculated by scaling upward the optical effect of the El Chichon eruption.

In Table 8.1 it can be seen that the silicate loading following the eruption of Mount St. Helens was dominated by large ash particles and the submicron mass was probably smaller than the 10^{13} g of submicron dust estimated for the baseline case. Likewise, the El Chichon silicate mass loadings consisted of large particles and the estimates are unreliable since no in situ measurements were made in the El Chichon cloud for several months, at which time the silicate dust loadings were very small (about 10^{12} g).

Unfortunately, the possibly large mass loadings of stratospheric submicron debris for prehistoric giant volcanic eruptions cannot be substantiated from current data. Very little study has been devoted to the environmental effects of giant volcanic eruptions. Clearly, the environmental effects may have been severe even thousands of kilometers from the volcanoes since substantial quantities of ash were deposited at such distances. The greatest volcanic eruptions, even if they do put as much as 10^{16} g into the atmosphere, would not be expected to cause effects similar to those of the much more powerful meteorite posited by Alvarez et al. as discussed by Toon et al. (1982). Major biological extinctions due to volcanic eruptions are then neither expected nor detected in the geologic record. Severe global climatic changes that would pose problems for modern society can neither be substantiated nor excluded on the basis of our current limited knowledge of prehistoric volcanic eruptions.

In summary, large explosive volcanic eruptions may be reasonable analogs for some atmospheric effects of a nuclear war, but not enough is known about these eruptions to provide useful guidelines. Clearly, studies of the amount of fine dust and aerosols that actually reaches the stratosphere are warranted. Likewise, a better knowledge of the environmental impact of previous eruptions is needed.

EXTRATERRESTRIAL IMPACTS

Alvarez et al. (1980) suggested that the abrupt extinction of many species of marine plankton and other organisms at the end of the Cretaceous period (about 65 million years ago) was a consequence of the impact of an extraterrestrial body of about 10 km in diameter that lofted quantities of cratering dust particles into the atmosphere. Subsequent work by Alvarez and others has led to the discovery at the Cretaceous-Tertiary boundary of a characteristic claystone layer enriched in certain noble metals at about 60 additional localities around the globe. Between one-fourth and one-half of the late Cretaceous plant taxa recognizable from pollen became extinct at the claystone layer, where it has been observed in North America. These new observations greatly strengthen the initial hypothesis of Alvarez et al. that a body of about 10 km in diameter did strike the earth at the end of the Cretaceous and that this impact may have caused the extinction of species. The extinction may have been a consequence of the darkening of most of the earth's surface by cratering debris suspended in the atmosphere (Alvarez et al., 1982; Toon et al., 1982), of an increase in surface temperature after the debris had settled due to an increase in the water vapor content of the atmosphere (Emiliani et al., 1981), of the production of large quantities of NO_x in the impact fireball (Lewis et al., 1982), or of the interaction of physical and biological effects. The suggestion that a global veil of dust was the cause of the extinction of species at the end of the Cretaceous spurred the present concern that the dust and soot produced in a nuclear war might have similarly deleterious effects.

Shoemaker (1983) estimates that the cumulative frequency of impacts by extraterrestrial bodies with radii $\geq r$ varies approximately as r^{-2}, at least for bodies in the kilometer size range. At a representative velocity of 20 km/s, the impact of a 10-km body would release an amount of energy of the order of 10^8 Mt. The Cretaceous-Tertiary claystone layer, where it is recognized, is about 2 cm thick on average, and the total mass of a global deposit with this thickness is about 10^{19} g. Impacts of bodies of 5-km diameter or larger occur about 4 times as frequently as impacts of 10-km bodies, or at a rate of 10^{-7} yr^{-1}, with impact energies of 10^7 Mt; for 2-km-diameter bodies or larger the rate is 6 x 10^{-7} yr^{-1} and the energy is 10^6 Mt. The masses of ejecta scale approximately with yield; scaling from the Cretaceous-Tertiary boundary claystone layer gives 10^{18} g and 10^{17} g (not necessarily all of it submicron sizes) globally dispersed ejecta for the 5-km and 2-km bodies, respectively. The resultant clay layers would be roughly 0.2 and 0.02 cm thick, respectively, and would not have been detected in the geologic record.

Although the particle size distribution in impact-generated dust clouds is unknown and the clouds could have had much smaller submicron fractions than the nuclear clouds, estimates of impact-produced dust could be from one to several orders of magnitude larger than the 2 x 10^{14} g of smoke and 2 x 10^{13} g of submicron dust generated in the baseline nuclear war. The impact energies are also orders of magnitude greater than the 6500-Mt yield of the baseline nuclear war. Impacts roughly comparable to the baseline war in energy release and in dust lofted require objects of about 500-m diameter, and occur roughly once every 10^5 yr.

Mass extinction events comparable to that at the end of the Cretaceous are fairly rare events in the history of the earth. Global mass extinctions that have been recognized at the taxonomic level of families of organisms have recurred at a mean interval of about 30 million years (Raup and Sepkoski, 1984). There is some evidence that mass extinctions might have been produced by the impact of bodies as small as 5 km in diameter, but the impact of 2-km bodies appears not to have left an easily recognized imprint on the succession of life forms recorded by fossils. It is clear that traceable catastrophes of the magnitude of the Cretaceous-Tertiary extinction are only produced by impacts with energy releases substantially exceeding those of a possible near-term global nuclear exchange. On the other hand, quite severe perturbations of the environment that did not succeed in producing extinction of many species--and with durations of a few years or less--cannot be easily detected in the stratigraphic record. Therefore it is not, at present, known whether impacts with energies in the range of 10^4 to 10^6 Mt had atmospheric effects similar to or even more severe than those projected for the baseline nuclear war.

Further studies of the environmental effects of large asteroidal impacts as well as studies of the debris lofted by asteroid impacts may be valuable for establishing analogs for the nuclear war case. Toon et al. (1982) have shown that impact events that produce masses of dust as small as 10^{17} g should have produced light levels and low temperatures very similar to those of 10^{19}-g impacts. Moreover, light levels low enough to cause failure of photosynthesis may occur with injections of no more than 10^{16} or 10^{17} g of dust. Hence the impacts of bodies 2 km in diameter or somewhat smaller may have produced both physical and biological effects that would be detectable in the stratigraphic record by means of a directed intensive search.

REFERENCES

Alvarez, L.W., F. Asaro, and H.V. Michel (1980) Extraterrestrial causes for the Cretaceous-Tertiary extinction. Science 208:1095-1108.

Alvarez, W.L., F. Asaro, and H.V. Michel (1982) Current status of the impact theory for the terminal Cretaceous extinction. Geol. Soc. Am. Spec. Pap. 190:305-315.

Cadle, R.D., C.S. Kiang, and J.F. Louis (1976) The global scale dispersion of the eruption clouds from major volcanic eruptions. J. Geophys. Res. 81:3125-3132.

Chuan, R.L., W.I. Rose, and D.C. Woods (1984) Size and chemistry of small particles in eruption clouds, in Clastic Particles: Scanning Electron Microscopy and Shape Analysis of Sedimentary and Volcanic Clasts, edited by J. Marshall. Stroudsburg, Pa.: Hutchinson Ross (in press).

Chuan, R.L., D.C. Woods, and M.P. McCormick (1981) Characterization of aerosols from eruptions of Mount St. Helens. Science 211:830-832.

Clanton, U.S., J.L. Gooding, and D.P. Blanchard (1982) Volcanic ash "clusters" in the stratosphere after the El Chichon, Mexico, eruption. Eos Trans. AGU 64:1139.

Deirmendjian, D. (1973) On volcanic and other particulate turbidity anomalies. J. Adv. Geophys. 16:267-297.

Emiliani, C., E.B. Kraus, and E.M. Shoemaker (1981) Sudden death at the end of Mesozoic. Earth Planet. Sci. Lett. 55:317-334.

Evans, W.F.J., and J.B. Kerr (1983) Estimates of the amount of sulfur dioxide injected into the stratosphere by the explosive volcanic eruptions--El Chichon, Mystery Vulcan, Mount St. Helens. Geophys. Res. Lett. 10:1049-1052.

Farlow, N.H., V.R. Overbeck, K.G. Snetsinger, G.V. Ferry, G. Polkowski, and D.M. Hayes (1981) Size distributions and mineralogy of ash particles in the stratosphere from eruptions of Mount St. Helens. Science 211:832-834.

Francis, P. (1983) Giant volcanic calderas. Sci. Am. 248(6):60-70.

Gooding, J.L., U.S. Clanton, E.M. Gabel, and J.L. Warren (1983) El Chichon volcanic ash in the stratosphere--Particle abundances and size distributions after the 1982 eruption. Geophys. Res. Lett. 10:1033-1036.

Hammer, C.U., H.B. Clausen, and W. Dansgaard (1980) Greenland ice sheet evidence of post-glacial volcanism and its climatic impact. Nature 228:230-235.

Hirono, M., and T. Shibata (1983) Enormous increase in stratospheric aerosols over Fukuoka due to volcanic eruption of El Chichon in 1982. Geophys. Res. Lett. 10:152-154.

Hofmann, D.J., and J.M. Rosen (1983) Stratospheric sulfuric acid fraction and mass estimate for the 1982 volcanic eruption of El Chichon. Geophys. Res. Lett. 10:313-316.

Humphreys, W.J. (1940) Physics of the Air. New York: McGraw-Hill.

Iwasaki, Y., S. Hayashida, and A. Ono (1983) Increasing backscattered light from the stratospheric aerosol layer after Mt. El Chichon eruption. Geophys. Res. Lett. 10:440-442.

Kennett, J.P., and R.C. Thunnel (1977) On explosive Cenozoic volcanism and climatic implication. Science 196:1231-1234.

Krueger, A.J. (1983) Sighting of El Chichon sulfur dioxide clouds with the Nimbus 7 total ozone mapping spectrometer. Science 220:1377-1379.

Landsberg, H.E., and J.M. Albert (1974) The summer of 1816 and volcanism. Weatherwise 27:63-66.

Lazrus, A.L., R.D. Cadle, B.W. Gandrud, J.P. Greenberg, B.J. Huebert, and W.I. Rose (1979) Sulfur and halogen chemistry of the stratosphere and of volcanic eruption plumes. J. Geophys. Res. 84:7869-7875.

Lewis, J.S., G.H. Watkins, H. Hartman, and R.G. Prinn (1982) Chemical consequences of major impact events on earth. Geol. Soc. Am. Spec. Pap. 190:215-221.

Mossop, S.C. (1963) Stratospheric particles at 20 km. Nature 199:325-327.

Mossop, S.C. (1965) Stratospheric particles at 20 km altitude. Geochim. Cosmochim. Acta 29:201-207.

Murrow, P.J., W.I. Rose, and S. Self (1980) Determination of the total grain size distribution in a volcanian eruption column, and its implications to stratospheric aerosol perturbation. Geophys. Res. Lett. 7:893-896.

Newell, R.E., and A. Deepak (1982) Mount St. Helens eruptions of 1980: Atmospheric effects and potential climatic impact. NASA SP-458. 119 pp.

Pollack, J.B., O.B. Toon, C. Sagan, A. Summers, B. Baldwin, and W. Van Camp (1976) Volcanic explosions and climatic change: A theoretical assessment. J. Geophys. Res. 81(6):1071-1083.

Rampino, M.R., and S. Self (1982) Historic eruptions of Tambora (1825), Krakatau (1853), and Agung (1963): Their stratospheric aerosols, and climatic impact. Quat. Res. 18:127-143.

Raup, D.M., and J.J. Sepkoski, Jr. (1984) Periodicity of extinctions in the geologic past. Proc. Nat. Acad. Sci. USA 81:801-805.

Robock, A. (1978) Internally and externally caused climate change. J. Atmos Sci. 35:1111-1122.

Rose, W.I., and M.F. Hoffman (1982) The May 18, 1980 eruption of Mount St. Helens: The nature of the eruption, with an atmospheric perspective. NASA CP-2240. Pages 1-14.

Rose, W.I., A.T. Anderson, S. Bonis, and L.G. Woodruff (1978) The October 1974 basaltic tephra from Fuego Volcano, Guatemala: Description and history of the magma body. J. Volcanol. Geotherm. Res. 4:3-53.

Rose, W.I., Jr., T.J. Bornhorst, S.P. Halsor, W.A. Capaul, and P.S. Plumley (1983a) Volcan El Chichon, Mexico: Pre-1982 S-rich eruptive activity. J. Volcanol. Geotherm. Res. 23:(in press).

Rose, W.I., R.L. Wunderman, M.F. Hoffman, and L. Gale (1983b) A volcanologist's review of atmospheric hazards of volcanic activity. J. Volcanol. Geotherm. Res. 17:133-157.

Sedlacek, W.A., E.J. Mroz, A.L. Lazrus, and B.W. Gandrud (1983) A decade of stratospheric sulfate measurements compared with observations of volcanic eruptions. J. Geophys. Res. 88:3741-3776.

Self, S., M.R. Rampino, and J.J. Barbera (1981) The possible effects of large 19th and 20th century volcanic eruptions on zonal and hemispheric surface temperatures. J. Volcanol. Geotherm. Res. 11:41-60.

Sepkoski, J.J., Jr. (1982) Mass extinctions in the Phaneozoic oceans: A review. Geol. Soc. Am. Spec. Pap. 190:283-289.

Settle, M. (1978) Volcanic eruption clouds and the thermal power output of explosive eruptions. J. Volcanol. Geotherm. Res. 3:309-324.

Shoemaker, E.M. (1983) Asteroid and comet bombardment of the earth. Ann. Rev. Earth Planet. Sci. 11:461-494.

Sigürdsson, H., S. Carey, and J.M. Espinbola (1984) 1982 eruptions of El Chichon Volcano, Mexico: Stratigraphy of pyroclastic deposits. J. Volcanol. Geotherm. Res. 23:(in press).

Simkin, T. (1981) Volcanoes of the World. Smithsonian Institution. Stroudsburg, Pa.: Hutchinson Ross.

Toon, O.B., and J.B. Pollock (1982) Stratospheric aerosols and climate. Pages 121-147 in The Stratospheric Aerosol Layer, edited by R.C. Whitten. Berlin: Springer Verlag.

Toon, O.B., J.B. Pollack, T.P. Ackerman, R.P. Turco, C.P. McKay, and M.S. Liu (1982) Evolution of an impact generated dust cloud and its effects on the atmosphere. Geol. Soc. Am. Spec. Pap. 190.

Varekamp, J.C., J. Luhr, and K. Prestegaard (1984) The 1982 eruptions of El Chichon Volcano (Chiapas, Mexico): Character of the eruptions, ash-fall deposits, and gas phase. J. Volcanol. Geotherm. Res. 23:(in press).

Wilson, L., R.S.J. Sparks, T.C. Huang, and N.D. Watkins (1978) The control of volcanic column heights by eruption energetics and dynamics. J. Geophys. Res. 83(B4):1829-1836.

APPENDIX: Evolution of Knowledge About Long-Term Nuclear Effects

The first nuclear explosion (the Trinity event) occurred in the desert near Alamogordo, New Mexico, on July 16, 1945. During the subsequent development of nuclear weapons, which has spanned four decades, the outcome of nuclear events have repeatedly impressed, and occasionally surprised, nuclear scientists and engineers.

The only two nuclear bombs to be used in wartime (detonated over Hiroshima and Nagasaki, Japan, in August 1945) each destroyed an entire city, although both were of quite low energy yield by today's standards. The 15-Mt Bravo test on Bikini Atoll in March 1954 underlined the hazard of radioactive fallout. The residents of Rongelap Atoll, more than 150 km downwind of Bikini, were exposed to, and suffered from, serious doses of nuclear fallout radiation even though they were quickly evacuated (Glasstone and Dolan, 1977). Following the first successful detonation of a fusion device in 1952, the pace of nuclear testing, the size of individual nuclear warheads, and the total nuclear arsenals of the United States and the USSR expanded rapidly (the USSR detonated the largest weapon, a ≈58-Mt device, in the atmosphere in October 1961).

Throughout this period and into the early 1960s, a debate developed among nuclear strategists as to whether blast or thermal (fire) effects should be considered the primary destruction mechanism in formulating nuclear strategy. Blast effects were finally settled on because they were certain to occur with each explosion; fire was considered a secondary effect, as was prompt radioactive fallout from surface bursts.

With the growth of the arsenals, scientists became concerned that severe global environmental effects might occur if even a fraction of the existing nuclear weapons were detonated. Such concern led, for example, to projects Gabriel and Sunshine--from 1949 through 1959--to evaluate the danger of radioactive fallout. Batten (1966) later assessed the possible climatic impact of dust raised by nuclear surface bursts, while Ayers (1965) undertook a broad analysis of the environmental and biological consequences of nuclear war, including the effects of blast, fires, and fallout. These early studies were hampered by a lack of critical data (some of which are now available) and were based on assumptions that seemed reasonable at the time but in retrospect appear to have been incorrect. The studies also were not quantitative in many important details. Ayers noted that very severe effects were possible in his scenarios, but he could not marshal the data to make a convincing case.

During the period of assessment of the global impacts of supersonic flight, Foley and Ruderman (1973) pointed out that the nitrogen oxides (NO_x) produced in nuclear fireballs by megaton-size explosions would be carried into the stratosphere. There NO_x would react with and deplete the ozone layer, which shields the earth from harmful ultraviolet sunlight. Hampson (1974) suggested that a full-scale superpower nuclear exchange could result in the nearly complete depletion of the ozone shield, possibly subjecting the earth to high levels of ultraviolet radiation for a year or more.

The 1975 National Research Council study (NRC, 1975) attempted to resolve some of these questions about the long-term effects of nuclear war. Much of that analysis centered on the recently identified ozone depletion problem. The report concluded that large reductions (about 50 percent) of the global ozone burden could occur. The NRC report judged that the likely climatic impact of nuclear dust from 10,000 Mt of high-yield surface explosions would probably be no more than the slight cooling produced by the great Krakatau eruption of 1883; but it noted a large uncertainty in these findings.

The recent renewal of interest in long-term effects of a nuclear war arose from two independent activities. One started in a seemingly unrelated field. Analysis of a thin clay layer found widely distributed at the stratigraphic boundary between the Cretaceous and Tertiary periods led Alvarez et al. (1980) to theorize--on the basis of anomalous levels of such noble metals as iridium--that the mass extinction of species that occurred 65 million years ago could be attributed to an asteroid striking the earth. The asteroid, they proposed, had raised a global dust cloud that blocked out sunlight so effectively that the terrestrial and marine food chains supporting the dinosaurs and many other species collapsed. Recognizing a possible parallel between the dust-lofting effect of an asteroid and that of a sizeable exchange of nuclear warheads, William J. Moran, in March 1981, stimulated discussions and preliminary calculations within the NRC. This work led to two meetings of an NRC study panel (December 1981 and April 1982) to further investigate the effects of dust lofted by nuclear detonations.

The second key event was the realization of the possible effects of smoke. Prior to this time, assessments of the effects of nuclear war did not include the potential effects of the smoke emitted by fires ignited by nuclear detonation. Attention to fires had focused instead on the immediate damage caused by burning and high temperatures. As part of a study that had been launched in 1980 by the Royal Swedish Academy of Sciences, Crutzen and Birks in early 1982 circulated a draft paper (published in June 1982) that provided the first quantitative evidence of the possible importance of smoke in blocking solar radiation, and suggested consequent alterations of weather and short-term climate in the northern hemisphere.

As an input to the April 1982 meeting, Turco, with the assistance of Toon, Pollack, and Ackerman--drawing upon the work in progress of Toon et al. (1982) on the climatic impact of dust lofted by an asteroidal impact--presented preliminary calculations on the climatic impact of nuclear dust emissions. The work of Crutzen and Birks on smoke was also reported by Turco and Eric Jones, and its potential importance was immediately recognized by the NRC study panel. A letter report (Moran, 1982) concluded that sufficient scientific data were available to warrant a thorough examination of the environmental effects of a nuclear exchange. Discussions then began between the NRC and the Department of Defense that culminated in the request for the present study. Meanwhile, Turco and his colleagues continued their studies presented at the April meeting and soon made the first quantitative estimates of the climatic effects of smoke and dust mixtures (Turco et al., 1982, 1983a,b). Related work by Crutzen et al. (1984), scientists at the Lawrence Livermore National Laboratory (e.g., MacCracken, 1983), and climatologists at the National Center for Atmospheric Research (Covey et al., 1984) underscored the potential seriousness of the problem.

Thus, nearly four decades after the introduction of nuclear weapons technology, a series of unplanned, separate scientific developments has led to a reevaluation of our understanding of the global effects of nuclear war. One can ask whether even now the full range of physical consequences--let alone the biological effects--of nuclear warfare is within our comprehension.

REFERENCES

Alvarez, L.W., W. Alvarez, F. Asaro, and H.W. Michel (1980) Extraterrestrial cause for the Cretaceous-Tertiary extinction. Science 208:1095-1108.

Ayers, R.U. (1965) Environmental Effects of Nuclear Weapons. Vols. 1-3. Report HI-518-RR. Harmon-on-Hudson, N.Y.: Hudson Institute.

Batten, E.S. (1966) The Effects of Nuclear War on the Weather and Climate. Memorandum RM-4989-TAB. Santa Monica, Calif.: RAND Corp. 50 pp.

Bethe, H. (1976) Ultimate catastrophe? Bull. At. Sci. 32:36-37.

Covey, C., S.H. Schneider, and S.L. Thompson (1984) Global atmospheric effects of massive smoke injections from a nuclear war: Results from general circulation model simulations. Nature 308:21-31.

Crutzen, P.J., and J.W. Birks (1982) The atmosphere after a nuclear war: Twilight at noon. Ambio 11:114-125.

Crutzen, P.J., C. Brühl, and I.E. Galbally (1984) Atmospheric effects from post-nuclear fires. Climatic Change, in press.

Foley, H.M., and M.A. Ruderman (1973) Stratospheric NO production from past nuclear explosions. J. Geophys. Res. 78:4441-4450.

Glasstone, S., and P.J. Dolan (eds.) (1977) The Effects of Nuclear Weapons. Washington, D.C.: U.S. Department of Defense. 653 pp.

Hampson, J. (1974) Photochemical war on the atmosphere. Nature 250:189-191.

MacCracken, M.C. (1983) Nuclear war: Preliminary estimates of the climatic effects of a nuclear exchange. Paper presented at the International Seminar on Nuclear War, 3rd Session: The Technical Basis for Peace. Ettore Majorana Centre for Scientific Culture, Erice, Sicily, Aug. 19-24, 1983.

Moran, W.J. (1982) Letter report to Frank Press, chairman of the National Research Council, April 20, 1982.

National Research Council (1975) Long-Term Worldwide Effects of Multiple Nuclear Weapons Detonations. Washington, D.C.: National Academy of Sciences.

Silver, L.T., and P.H. Schultz (eds.) (1982) Geological implications of impacts of large asteroids and comets on the earth. Geol. Soc. Am. Spec. Pap. 190. 328 pp.

Toon, O.B., J.B. Pollack, T.P. Ackerman, R.P. Turco, C.P. McKay, and M.S. Liu (1982) Evolution of an impact generated dust cloud and its effects on the atmosphere. Geol. Soc. Am. Spec. Pap. 190:187-200.

Turco, R.P., O.B. Toon, J.B. Pollack, and C. Sagan (1982) Global consequences of nuclear warfare. Eos Trans. AGU 63:1018.

Turco, R.P., O.B. Toon, T.P. Ackerman, J.B. Pollack, and C. Sagan (1983a) Nuclear winter: Global consequences of multiple nuclear explosions. Science 222:1283-1292.

Turco, R.P., O.B. Toon, T.P. Ackerman, J.B. Pollack, and C. Sagan (1983b) Global Atmospheric Consequences of Nuclear War. Interim Report. Marina del Rey, Calif.: R&D Associates. 144 pp.

Index

G

H

I

L

M

N

O

P

V

W